Battery-Powered Electric and Hybrid Electric Vehicle Projects to Reduce Greenhouse Gas Emissions:

A Resource Guide for Project Development

July 2002

Prepared for the National Energy Technology Laboratory (NETL)
626 Cochrans Mill Road
P.O. Box 10940
Pittsburgh, PA 15236-0940
www.netl.doe.gov

by

Science Applications International Corporation (SAIC)
Climate Change Services
8301 Greensboro Drive, E-5-7
McLean, Virginia 22102
www.saic.com

With contributions from:
Orestes Anastasia, Nancy Checklick, Vivianne Couts, Julie Doherty,
Jette Findsen, Laura Gehlin, and Josh Radoff
SAIC

Reviewed by:
Richard Bechtold, QSS, Inc.
Eric Bell, NETL
Melissa Chan, NETL
Jim Ekmann, NETL
George Lee, NETL

Disclaimer

Table of Contents

Acknowledgements

The authors express their appreciation to Jim Ekmann of the National Energy Technology Laboratory for his ideas, inspiration and support. We also thank Rich Bechtold, Eric Bell, Melissa Chan, and George Lee who provided insightful review comments and helped refine some of the material presented here. Finally, we thank Marcy Rood of the U.S. Department of Energy's Clean Cities Program for her encouragement and support for this effort.

Acronyms Used in this Report

ACP	Alternative Compliance Plan
AFV	Alternative Fuel Vehicle
AIJ	Activities Implemented Jointly
AT PZEV	Advanced Technology Partial Zero-Emission Vehicle
BEV	Battery-Powered Electric Vehicles
CAA	Federal Clean Air Act
CAFÉ	Corporate Average Fuel Economy
CARB	California Air Resources Board
CARFG2	California Phase 2 Reformulated Gasoline
CCX	The Chicago Climate Exchange
CD	Conventional Diesel
CDM	Clean Development Mechanism
CEC	California Energy Commission
CH_4	Methane
CIDI	Compression Ignition, Direct Injection
CO	Carbon Monoxide
CO_2	Carbon Dioxide
CO_2E	Carbon Dioxide Equivalent
CNG	Compressed Natural Gas
CSDA	Center for Sustainable Development in the Americas
CV	Conventional Vehicle
DOE	U.S. Department of Energy
E10	A Mixture of 10% Ethanol and 90% Gasoline
E85	A Mixture of 85% Ethanol and 15% Gasoline
EIA	Energy Information Administration
EPA	Environmental Protection Agency
EPAct	Energy Policy Act of 1992
E.O.	Executive Order
ERUPT	Emission Reduction Unit Procurement Tender, The Netherlands
ETBE	Ethyl Tertiary Butyl Ether
EV	Electric Vehicle
FFV	Fuel Flexible Vehicle
GHG	Greenhouse Gas
GI	Grid Independent
GREET	Greenhouse Gases, Regulated Emissions, and Energy Use in Transportation
GV	Gasoline Vehicle
GVW	Gross Vehicle Weight
GWP	Global Warming Potential
HC	Hydrocarbon
HEV	Hybrid-Electric Vehicle
HOV	High-Occupancy Vehicle
ICE	Internal Combustion Engine
IPCC	Intergovernmental Panel on Climate Change
IRS	Internal Revenue Service
LDT	Light-Duty Truck
LEV	Low-Emission Vehicle
LNG	Liquid Natural Gas
LPG	Liquid Petroleum Gas
Mpg	Miles per gallon
MTBE	Methyl Tertiary Butyl Ether
MY	Model Year

M85	A Mixture of 85% Methanol and 15% Gasoline
NETL	National Energy Technology Laboratory
NEV	Neighborhood Electric Vehicle
NGV	Natural Gas Vehicle
NMOG	Non-Methane Organic Gas Standard
NO_X	Nitrous Oxides (unspecified)
N_2O	Nitrous Oxide
OEM	Original Equipment Manufacturer
PC	Passenger Car
PCF	Prototype Carbon Fund
PEF	Petroleum Equivalency Factor
PERT	Pilot Emissions Reduction Program, Canada
P.L.	Public Law
PM	Particulate Matter
PZEV	Partial Zero-Emission Vehicle
RFG	Reformulated Gasoline
RPE	Retail Price Equivalent
SIDI	Spark Ignition, Direct Injection
SULEV	Super-Ultra-Low-Emission Vehicle
SUV	Sport Utility Vehicle
TLEV	Transitional Low Emissions Vehicle
ULEV	Ultra-Low-Emission Vehicle
USIJI	U.S. Initiative on Joint Implementation
UNFCCC	United Nations Framework Convention on Climate Change
WBCSD	World Business Council for Sustainable Development
WRI	World Resources Institute
ZEV	Zero-Emission Vehicle

[this page deliberately left blank]

➡ Overview

The transportation sector accounts for a large and growing share of global greenhouse gas (GHG) emissions. Worldwide, motor vehicles emit well over 900 million metric tons of carbon dioxide (CO_2) each year, accounting for more than 15 percent of global fossil fuel-derived CO_2 emissions.[1] In the industrialized world alone, 20-25 percent of GHG emissions come from the transportation sector. The share of transport-related emissions is growing rapidly due to the continued increase in transportation activity.[2] In 1950, there were only 70 million cars, trucks, and buses on the world's roads. By 1994, there were about nine times that number, or 630 million vehicles. Since the early 1970s, the global fleet has been growing at a rate of 16 million vehicles per year. This expansion has been accompanied by a similar growth in fuel consumption.[3] If this kind of linear growth continues, by the year 2025 there will be well over one billion vehicles on the world's roads.[4]

In a response to the significant growth in transportation-related GHG emissions, governments and policy makers worldwide are considering methods to reverse this trend. However, due to the particular make-up of the transportation sector, regulating and reducing emissions from this sector poses a significant challenge. Unlike stationary fuel combustion, transportation-related emissions come from dispersed sources. Only a few point-source emitters, such as oil/natural gas wells, refineries, or compressor stations, contribute to emissions from the transportation sector. The majority of transport-related emissions come from the millions of vehicles traveling the world's roads. As a result, successful GHG mitigation policies must find ways to target all of these small, non-point source emitters, either through regulatory means or through various incentive programs. To increase their effectiveness, policies to control emissions from the transportation sector often utilize indirect means to reduce emissions, such as requiring specific technology improvements or an increase in fuel efficiency. Site-specific project activities can also be undertaken to help decrease GHG emissions, although the use of such measures is less common. Sample activities include switching to less GHG-intensive vehicle options, such as electric vehicles (EVs) or hybrid electric vehicles (HEVs). As emissions from transportation activities continue to rise, it will be necessary to promote both types of abatement activities in order to reverse the current emissions path. This Resource Guide focuses on site- and project-specific transportation activities.

Over the last decade, efforts to reduce GHG emissions in the U.S. have led to the creation of a number of voluntary programs for registering and crediting project-specific

---

[1] World Resources Institute, *Proceed With Caution: Growth in the Global Motor Vehicle Fleet,* http://www.wri.org/trends/autos.html.

[2] "Good Practice Greenhouse Abatement Policies: Transport Sector," OECD and EIA Information Papers prepared for the Annex I Expert Group on the UNFCCC. OECD and IEA, Paris, November 2000. Emissions exclude land-use change and forestry, and bunker fuels. Annex I countries are those countries that have undertaken binding emission reduction targets under the Kyoto Protocol of the United Nations Framework Convention on Climate Change (UNFCCC).

[3] American Automobile Manufacturers Association (AAMA), *World Motor Vehicle Data 1993* (AAMA, Washington, D.C., 1993), p. 23, and American Automobile Manufacturers Association (AAMA), *Motor Vehicle Facts and Figures 1996* (AAMA, Washington, D.C., 1996).

[4] World Resources Institute, *Proceed With Caution: Growth in the Global Motor Vehicle Fleet,* http://www.wri.org/trends/autos.html.

GHG reduction activities undertaken by individual project developers. Similarly, several international programs have been implemented, including efforts that allow for trading in GHG emission reduction activities. As a result, a small but growing market for the trade in GHG emission reduction credits has emerged, creating an additional incentive for project developers in the transportation sector to undertake GHG reduction projects. Given that EVs and HEVs both emit less GHG emissions compared to conventional vehicles, projects that lead to the introduction of EVs and HEVs could register with the many voluntary GHG reporting programs and may be able to sell the associated GHG reduction credits on the market. However, to participate in these efforts, project developers must be familiar with the procedures for developing and estimating the GHG emissions benefits resulting from the various types of projects.

This National Energy Technology Laboratory (NETL) publication, "*Battery-Powered Electric and Hybrid Electric Vehicles to Reduce Greenhouse Gas (GHG) Emissions: A Resource Guide for Project Development*" provides national and international project developers with a guide on how to estimate and document the GHG emission reduction benefits and/or penalties of battery-powered and hybrid-electric vehicle projects. This primer also provides a resource for the creation of GHG emission reduction projects for the Activities Implemented Jointly (AIJ) Pilot Phase and in anticipation of other market-based project mechanisms proposed under the United Nations Framework Convention on Climate Change (UNFCCC). Though it will be necessary for project developers and other entities to evaluate the emission benefits of each project on a case-by-case basis, this primer will provide a guide for determining which data and information to include during the process of developing the project proposal.

The resource guide first provides an overview of the various technology options for both EVs and HEVs. Sections 1 and 2 briefly summarize the types of EVs and HEVs available and review their performance and estimated costs. These introductory sections are followed by Section 3, which provides an overview of the emerging regulatory frameworks promoting the use of EV and HEVs, including relevant domestic and international climate change policy developments. Section 4 discusses the procedures for estimating GHG emission reductions from EV and HEV projects. This includes a summary of the GHG emissions associated with EVs and HEVs, an overview of studies analyzing potential climate change-related benefits, a description of EV and HEV projects previously implemented, and a discussion of the common procedures required to participate in project-based GHG reduction systems. Finally, section 5 presents a hypothetical case study on how to develop a baseline and estimate the resulting GHG benefits from an EV or HEV project.

1   Hybrid Electric Vehicle Technology Options

HEVs combine the internal combustion engine of a conventional vehicle with the battery and electric motor of an electric vehicle. This combination offers the extended range and rapid refueling that consumers expect from a conventional vehicle, with a significant portion of the energy and environmental benefits of an electric vehicle. The practical benefits of HEVs include improved fuel economy and lower emissions of the full host of criteria pollutants, as well as CO_2, compared to conventional vehicles. The inherent flexibility of HEVs will allow them to be used in a wide range of applications, where electric-only vehicles cannot, from personal and public transportation to commercial hauling. As with electric vehicles and conventional vehicles, the main factors that can be used to rate one vehicle over another are:

- Range: how far can the vehicle travel between refueling?
- Refueling time: how long does it take to refuel?
- Refueling infrastructure: how available are refueling stations?
- Efficiency or fuel economy or gas mileage: how far can the vehicle travel for a given unit of fuel energy, measured in miles per gallon for HEVs and conventional vehicles?
- Performance: how well does the vehicle handle?
- Power: what acceleration can the engine deliver? What speeds can it maintain?
- Safety: how vulnerable is the vehicle to collision? How quickly can it brake? and
- Cost.

These are the elements and questions to keep in mind when considering the relative merits of one vehicle type over another. In general, HEVs typically compare well with their conventional counterparts. They have increased range and mileage, and comparable power, safety and performance. Their main hindrance to mass commercial deployment has been their cost, but this too is becoming more and more competitive. The following paragraphs briefly discuss some of the basic characteristics of HEVs, and more detailed specifications can be found in Appendix 1.

1.1   Types of Vehicles

Many configurations are possible for HEVs. Essentially, a hybrid combines an energy storage system, a power unit, an electric motor, and a vehicle propulsion system. The primary options for energy storage include **batteries**, **ultracapacitors**, and **flywheels**. Although batteries are by far the most common energy storage choice, other possibilities are being researched. Hybrid power units are typically spark ignition internal combustion engines (similar to those employed in conventional vehicles) for light-duty hybrid vehicles and diesel engines for heavy-duty hybrids. Other "power plant" options for hybrids include spark ignition direct injection engines (SIDI), **gas turbines**, and **fuel cells**.

There are two main system types for propulsion, which define the two main branches of hybrid-electric technology:

- The series configuration, in which the propulsion comes solely from an electric **motor**. In this case the engine is used to continually repower the battery, and

- The parallel configuration, in which the role of the engine is to provide direct mechanical input to drive the vehicle in parallel with the electric motor, and also to charge the battery.

A hybrid's efficiency and emissions depend on the particular combination of subsystems, how these subsystems are integrated into a complete system, and the control strategy that integrates the subsystems. Hybrid vehicle fuel economy ranges from 10 to 15 percent higher than conventional vehicles (for "mild hybrids") to between 200 and 300 percent higher for the most advanced systems where increase in fuel efficiency is optimized. The potential gains in fuel efficiency by hybrids are dependent on the type of driving the vehicle is typically used for—higher gains are possible in congested urban driving than highway driving.

1.1.1 Commercially Available Vehicles

Currently, there are only two original equipment manufacturers (OEM) with HEVs on the U.S. market: Honda and Toyota (Honda has two, the Insight and a new Civic Hybrid, and Toyota has one, the Prius).

The Honda Insight was the first HEV to be available for public purchase – the two-seat model was introduced across the country in late 1999. The Insight uses an Integrated Motor Assist (IMA^{TM}) system, which combines the world's lightest 1.0-liter, 3-cylinder gasoline automobile engine with an ultra-thin electric motor. The U.S. Environmental Protection Agency (EPA) has rated the Insight as having a gas mileage of 61-mpg city/70-mpg highway. The cost of the Insight is around $19,000.

Honda Insight, 2002

The Prius was introduced shortly after the Insight and seen far more success in terms of overall sales, although still limited to about 2000 per month[5]. The main difference is that the Prius seats four and has the exterior of a standard sedan, which makes it more or less interchangeable with conventional passenger vehicles and makes users feel less like they are driving a concept car wherein performance and/or safety might be unreliable. As a result of the increased space and weight, the Prius' gas mileage is lower than that of the Insight, with an EPA rating of 52-mpg city/45-mpg highway. However, the success of the Prius, and the prospect for hybrids in general (especially those that are interchangeable with larger conventional vehicles), have prompted Honda to release a new hybrid model of its popular Civic. It too is a four-door sedan, and gets lower mileage

---

[5] Washington Post Article, "Half Gas, Half Electric, Total California Cool; Hollywood Gets a Charge Out of Hybrid Cars." June 6, 2002.

than the Insight at 51-mpg city/46 mpg highway, but Honda expects the sales to be sharply increased over the Insight and to rival or surpass those of the Prius. The cost of both the Prius and Civic hybrid are about $20,000. In the near future, additional HEVs, including Sport Utility HEVs and other large models, are expected from General Motors, Ford, DaimlerChrysler, and others.

Toyota Prius, 2002

1.2 Vehicle Performance

As mentioned previously, well-designed hybrid-electric vehicles have similar capabilities with regard to speed, safety and handling compared to conventional vehicles. In fact, hybrids have the same or greater range than traditional internal combustion engine vehicles – Honda's Insight can go about 700 miles on a single tank of gas, while the Toyota Prius can go about 500 miles – and achieve about twice the fuel economy of their conventional counterparts. The following table compares the characteristics of the Toyota Prius with a conventional automobile of a similar size class: the Honda Civic Sedan DX. For more detail, refer to Appendix 1.

Table 1-1 HEV Versus Conventional Vehicle Comparison[6]		
	Prius	**Honda Civic DX**
Power	98 Hp (combined engine and motor)	115 Hp @ 6,100 rpm
Maximum Speed	100 mph	108 mph
Acceleration	0-60 mph in 12.7 sec	0-60 mph in 10.2 seconds
Braking	front disk/rear drum with integrated regenerative system, anti-lock braking system	front disk/rear drum, anti-lock braking system
Fuel Efficiency	52 mpg city/45 mpg highway	30 mpg (city); 38 mpg (highway)
Emissions Rating	SULEV	ULEV
Range	619 miles city; 535 miles highway	396 miles city; 502 miles highway
Cost	~$20,000	~14,000

---

[6] Vehicle costs and characteristics were taken directly from the company websites. May-June, 2002.

Anecdotally, in terms of operating a new type of vehicle, some drivers have reported needing some time to get used to the fact that the energy-saving engines of some HEVs are designed to shut off automatically when the vehicle is braking or stopped at a red light.

2 Battery-Powered Electric Vehicle Technology Options

Battery operated electric vehicles are powered by an electric motor that draws on stored electricity from the on-board batteries. Battery operated electric vehicles are sometimes referred to as "zero emission vehicles" (ZEVs) (and are classified as such under certain regulatory regimes), because there are no tailpipe emissions (there is no tailpipe), nor are there emissions associated with fuel evaporation, refining, or transport. The designation of ZEV can be misleading, as there are a number of indirect emissions that can be associated with the vehicle—namely during the production of the electricity at the power plant. However, the fact that no emissions come from the vehicle itself has enormous significance in the context of urban air pollution and issues associated with the usage of gasoline and diesel, such as fuel security.

2.1 Types of Vehicles

EVs come in two basic types: "full function" EVs that are the equivalent of conventional light-duty vehicles, and "neighborhood EVs" that are small, typically two-seater vehicles with limited speed (top speed of 25 mph) and operating range (typically 30 miles) intended for use for short trips on non-highway roads. Neighborhood vehicles, which are recharged at the users' home, can satisfy short-trip transportation needs on local community roads that are not major thoroughfares. It should be noted that neighborhood EVs are not classified as or considered appropriate for use as typical passenger vehicles, nor do they meet the same safety standards as full function EVs and conventional vehicles. However, to date, 34 U.S. states and the District of Columbia have authorized their use on roads with 35 mph speed limits, for which there is a significant niche market.

The full function light-duty EVs offered for sale include two two-seater cars, one mid-size station wagon, one pickup truck, one small SUV, and a service van. Ford offers its Th!nk two-seater that has an operating range of 53 miles. General Motors offers their EV1, which has an operating range of up to 90 miles when lead-acid batteries are used, and up to 130 miles when nickel-metal hydride batteries are used. Nissan offers an electric version of their Altra station wagon to customers in California only. The Altra has an operating range of 80 miles using lithium-ion batteries. Ford has an electric version of their small pickup (Ranger) that has a range of 73 miles using lead-acid batteries. Toyota has an electric version of their RAV4 small SUV with an operating range of 126 miles using nickel-metal hydride batteries. The RAV4-EV is only sold to fleet customers in California.

2.2 Vehicle Performance

The driving ranges of battery-operated electric vehicles typically vary from 50 to 130 miles, depending on a vehicle's weight, its design features, and the type of battery it uses.[7] Drivers can refuel a battery-operated vehicle by simply plugging it into a special recharging outlet at home, which is both convenient in the sense of allowing drivers to

---

[7] "Just the Basics: Electric Vehicles, Transportation for the 21st Century," Office of Energy Efficiency and Renewable Energy, Office of Transportation Technologies, U.S. Department of Energy.

refuel overnight at home, and inconvenient, due to the fact that it can take extended periods of time to charge the vehicle. The recharging time depends on the voltage of the recharging station, the ambient air temperature, the size and type of the battery pack, and the remaining electrical energy in storage. Typically, the process takes several hours, but batteries are being developed that can be recharged more quickly.

Electric vehicles can be more efficient than conventional vehicles on a purely fuel to motive energy conversion basis, because electric motors are more efficient at low speeds unlike internal combustion engines, and because they do not use any power when coasting or at rest.[8] Adding to the efficiency of battery-powered electric vehicles is the technique of regenerative braking. Regenerative braking is the process of slowing and stopping a vehicle by converting its mechanical energy to electric energy, which can then be returned to the vehicle's on-board battery. In a conventional vehicle, this energy is simply wasted as heat. (Many hybrid vehicles also incorporate regenerative braking which contributes significantly to their efficiency improvements.) Typical figures for electric vehicle efficiencies range from 0.2 to 0.4 kWh per mile traveled.

To compare the GHG emissions from an electric vehicle to that of a conventional vehicle, one must first know the source of the electricity used to power the electric vehicle. Electricity derived from hydropower, wind power or other renewable resources would have no GHG emissions, while electricity derived from a coal plant would have nearly identical CO_2 emissions as that of a conventional 26 mpg gasoline passenger car[9]. Therefore a conventional car with above-average efficiency, say of 30 mpg or better, will produce less CO_2 than an EV powered by coal-derived electricity. On the other hand, an EV using electricity produced from a source other than coal, such as natural gas, will produce less CO_2 emissions. This topic is discussed further in section 4.

2.3 Vehicle Costs

Electric vehicles are about twice as expensive as their conventional fuel counterparts. For example, the EV1 is advertised to cost about $40,000. However, all the major auto manufacturers require that their EVs be leased instead of bought, since manufacturers are uncomfortable selling them at this time, given relative inexperience with maintenance, service, and recharging. The Altra wagon, RAV4-EV, and Ford Ranger EV all lease for $599 per month. Neighborhood electric vehicles range in price from $6,000 to as much as $20,000 depending on amenities and battery technology. For a sample study, completed by Argonne National Laboratory, analyzing the lifecycle costs of EVs versus conventional vehicles please refer to Appendix 4.

---

[8] It should be noted that when the electricity to be used in an EV is generated inefficiently, the resulting electric vehicle efficiency can often be worse than that of a conventional vehicle.

[9] Based on the following assumptions: electric vehicle efficiency of 0.3 kWh/mile (various sources report a range from 0.2 to 0.4 kWh/mile); gasoline vehicle efficiency of 26 miles per gallon; heat rate at coal plant of 9,750 BTU/KWH (35% efficiency); coal carbon content of 26.8 kg Carbon/GJ (IPCC); gasoline carbon content of 2.42 kg C/gallon (EIA).

3 Regulatory and Policy Frameworks Promoting Electric and Hybrid Electric Vehicles

Over the past decade, a number of regulatory policies have been introduced in the U.S. to promote the use of EVs and HEVs. Many of these policies are directed toward addressing urban air pollution and reducing fuel use, and not directly geared toward the reduction of GHG emissions. However, by promoting the adoption of more alternative fuel vehicles (AFVs),[10] such as EVs and HEVs, they indirectly contribute to the goal of reducing GHG emissions from the transportation sector. In the following discussion, this chapter considers a number of relevant regulations, policies, and programs that encourage the adoption and procurement of EVs and HEVs.

A wide variety of direct and indirect regulatory and policy drivers work to promote the broader use of AFVs in the United States, and encourage the development of EV/HEV projects. These measures typically include the following:

- Regulatory incentives, such as a tax credit or deduction, for the purchase or government procurement of an AFV or AFV-related equipment;
- Mandates and directives to both public and private fleet operators to purchase AFVs;
- Emissions standards for AFVs, low emission, or zero emission vehicles that encourage market shifts towards increased development and deployment of AFVs in the automobile market;
- Special regulations that provide advantages to AFV owners, such as access to high occupancy vehicle lanes, parking lanes, or simplified vehicle registration; and
- Fuel economy requirements for conventional vehicles that encourage development of more fuel efficiency technologies and vehicles, such as AFVs.

Each of these policy approaches has varying impacts on the promotion of AFVs. The chapter addresses federal and state regulatory policies that more directly promote the adoption of EVs and HEVs, primarily considering incentives, procurement mandates, and emissions and technology standards. This chapter is structured to first include an overview of Federal policies and programs targeting EVs and HEVs. The discussion begins with an overview of federal policy, primarily including tax incentives and AFV procurement mandates. Next, the discussion addresses major state policies and programs introduced to encourage the use of EV and HEVs, including California's low

---

[10] The term "alternative fueled vehicle" is defined as any dedicated vehicle or a dual fueled vehicle. (EPAct §301.) As provided in EPAct, the term "alternative fuel" is defined as:

> methanol, denatured ethanol, and other alcohols; mixtures containing 85 percent or more (or such other percentage, but not less than 70 percent, as determined by the Secretary, by rule, to provide for requirements relating to cold start, safety, or vehicle functions) by volume of methanol, denatured ethanol, and other alcohols with gasoline or other fuels; natural gas; liquefied petroleum gas; hydrogen; coal-derived liquid fuels; fuels (other than alcohol) derived from biological materials; electricity (including electricity from solar energy); and any other fuel the Secretary determines, by rule, is substantially not petroleum and would yield substantial energy security benefits and substantial environmental benefits.

emission and zero emission vehicle regulations. Finally, this section concludes with a discussion of the relevant *domestic and international climate change* policies and programs that are likely to have an influence on the development of EV and HEV markets. It should be noted that while gas and diesel HEV technologies are considered to be AFVs in certain contexts, neither are eligible for the majority of the Federal and state incentives presented below, and therefore the majority of the discussion with regard to incentives will pertain to EVs only. On the other hand, both EVs and HEVs figure prominently in some of the emerging regulations.

3.1 Federal Policies and Programs

Many of the most relevant elements of Federal policy to promote the development and use of alternative fuels in the transportation sector were introduced with the passage of the Energy Policy Act of 1992 (EPAct).[11] The primary motivations behind promoting alternative fuels under EPAct included reducing the nation's dependence on foreign oil and increasing the nation's energy security through the use of domestically produced alternative fuels. To do so, EPAct established a goal of replacing 10 percent of petroleum-based motor fuels in the United States by the year 2000 and 30 percent by the year 2010. As discussed below, EPAct addresses EVs and HEVs in two principal ways: first, by providing tax credits and deductions for the purchase of EVs and development of EV infrastructure, and second, by mandating Federal, State, and private "alternative fuel provider"[12] fleets to purchase AFVs (excluding HEVs).[13]

3.1.1 Federal Tax Incentives for Electric Vehicles

The Federal government introduced two forms of tax incentives relating to EVs under EPAct:

- a Federal tax credit available to individuals and businesses purchasing qualified EVs;

---

[11] Energy Policy Act of 1992, Public Law 102-486.

[12] According to the U.S. Department of Energy, an alternative fuel provider is defined as:
   □ [an entity] that owns, operates, leases, or otherwise controls 50 or more light-duty vehicles (LDVs) in the U.S. that are not on the list of EPAct Excluded Vehicles [such as emergency or law enforcement vehicles];
   □ least 20 of those LDVs are used primarily within a Metropolitan Statistical Area (MSA)/Consolidated Metropolitan Statistical Area (CMSA);
   □ those same 20 LDVs are centrally fueled or capable of being centrally fueled. LDVs are centrally fueled if they capable of being refueled at least 75% of the time at a location that is owned, operated, or controlled by any fleet, or under contract with that fleet for refueling purposes.
An alternative fuel provider is covered under EPAct if its principal business involves one of the following:
   □ producing, storing, refining, processing, transporting, distributing, importing, or selling any alternative fuel (other than electricity) at wholesale or retail;
   □ generating, transmitting, importing, or selling electricity at wholesale or retail; or
   □ produces and/or imports an average of 50,000 barrels per day or more of petroleum, as well as 30% or more of its gross annual revenues are derived from producing alternative fuels.
http://www.ott.doe.gov/epact/alt_fuel_prov.shtml.

[13] Although HEVs are fuel efficient and produce low levels of emissions, they do not count as "alternative fuel vehicles" because the HEVs on the market today use gasoline rather than alternative fuels. There has been some discussion on including HEVs as AFVs in the future, but no final decision has been made to date. http://www.ott.doe.gov/hev/faqs.html. See also http://www.ott.doe.gov/legislation.shtml.

- a Federal tax deduction for business expenses related to the incremental cost to purchase or convert to qualified clean fuel vehicles.

Electric Vehicle Tax Credit

EPAct established a tax credit for individuals or businesses purchasing qualified EVs that have been put into service between December 20, 1993 and December 31, 2004.[14] IRS Form 8834 can be used to determine the credit for qualified electric vehicles placed in service during the year. The tax credit is 10 percent of the purchase price, up to a maximum of $4,000, for qualified EVs placed in service before 2002. Beginning in 2001, the size of the credit is reduced by 25 percent of the original amount per year until the credit is fully phased out. Thus, the tax credit for each vehicle placed in service during 2002 is 7.5 percent of the cost of the qualified EV, up to a maximum credit of $3,000. Credits will be reduced by 50 percent for 2003 vehicles and by 75 percent for 2004 vehicles (see Table 3.1).

Table 3-1	Summary of Tax Credits for Qualifying Electric Vehicles
Date	**Deduction Available**
Dec. 20, 1993 - 2001	up to $4,000
2002	up to $3,000
2003	up to $2,000
2004	up to $1,000
2005	None - credit fully phased out

A qualified EV is defined as any motor vehicle that is powered primarily by an electric motor drawing current from rechargeable batteries, fuel cells, or other portable sources of electrical current, was manufactured primarily for use on public streets, roads, and highways, and has at least four wheels. All dedicated, plug-in-only EVs qualify for the tax credit. The credit does not apply to vehicles primarily used outside the United States, vehicles used by any governmental body or agency or any foreign person or entity, or vehicles used by a tax-exempt organization.[15] All series and some parallel HEVs meet the aforementioned qualifications, although HEVs that are not powered *primarily* by an electric motor, such as the Honda Insight or Toyota Prius, do not qualify as EVs. However, part of the cost of these parallel HEVs (up to $2,000 for a vehicle with a gross vehicle weight rating that does not exceed 10,000 pounds) may qualify for the deduction for clean-fuel vehicles, even if they are not used for business purposes.

Clean Fuel Vehicle Deduction

Similar to the EV tax credit, EPAct established a tax deduction for the purchase of a new OEM qualified clean fuel vehicle, or for the conversion of a vehicle to use a clean-burning fuel.[16] EPAct made available a Federal income tax deduction of between $2,000 and $50,000 (per vehicle) for the incremental cost to purchase or convert qualified clean fuel vehicles, including EVs, and for certain kinds of refueling property (see below).

The deductions are available for vehicles put into service between December 20, 1993 and December 31, 2004. Like the EV tax credit, the deduction will be reduced by 25

---

[14] EPAct, Title XIX-Revenue Provisions, Sec. 30, Credit for Qualified Electric Vehicles.

[15] *Alternative Fuel Vehicle Fleet Buyer's Guide,* http://www.fleets.doe.gov/cgi-bin/fleet/main.cgi?17357,state_ins_rep,5,468050; See also IRS 2001 Form 8834; see also IRS Publication 535.

[16] Public Law-102-486, Title XIX-Revenue Provisions, Sec. 179A.

percent of the original amount each year starting in 2001, and will be phased out completely by 2005. The amount of the tax deductions for qualified clean fuel vehicles is based on the gross vehicle weight and type of vehicle. The tax deduction for clean fuel vehicles is available for any applicable business or personal vehicle, except EVs eligible for the federal EV tax credit. The deduction is not amortized and must be taken in the year the vehicle is acquired.[17]

As provided in Table 3-2, the tax deduction for trucks or vans with gross vehicle weight of between 10,000 and 26,000 lbs is $5,000 per vehicle. The deduction is $50,000 per vehicle for trucks and vans over 26,000 lbs, or buses with seating capacity of 20 or more adults. Other clean fuel vehicles may qualify for a $2,000 credit. Table 3-2 also provides the maximum deductions for vehicles put into service after 2001 and through 2004, the final year the deduction may be taken before it is fully phased out.

Table 3-2. Summary of Deductions for Clean Fuel Vehicles

Date Vehicle Acquired	Vehicle Type	Deduction Available
Dec. 20, 1993 - 2001	truck or van with GVW 10,000-26,000 lbs.	$5,000
	truck or van with GVW over 26,000 lbs.	$50,000
	each bus, with seating capacity of at least 20 adults (excluding driver)	$50,000
	all other vehicles (excluding off-road vehicles)	$2,000
2002	truck or van with GVW 10,000-26,000 lbs.	$3,750
	truck or van with GVW over 26,000 lbs.	$37,500
	each bus, with seating capacity of at least 20 adults (excluding driver)	$37,500
	all other vehicles (excluding off-road vehicles)	$1,500
2003	truck or van with GVW 10,000-26,000 lbs.	$2,500
	truck or van with GVW over 26,000 lbs.	$25,000
	each bus, with seating capacity of at least 20 adults (excluding the driver)	$25,000
	all other vehicles (excluding off-road vehicles)	$1,000
2004	truck or van with GVW 10,000-26,000 lbs.	$1,250
	truck or van with GVW over 26,000 lbs.	$12,500
	each bus, with seating capacity of at least 20 adults (excluding the driver)	$12,500
	all other vehicles (excluding off-road vehicles)	$500
2005	all vehicles	None—deduction fully phased out

Deduction for EV Recharging Property

A tax deduction of up to $100,000 is also available for qualified "recharging property" for EVs being used in a trade or business, per location. Recharging property includes any equipment used to charge the electric battery of motor vehicle propelled by electricity and includes: low-voltage and high-voltage recharging equipment, quick-charging equipment, and ancillary connection equipment such as inductive chargers. It also includes the battery itself.[18]

---

[17] U.S. Department of Energy, *Alternative Fuel Vehicle Fleet Buyer's Guide*, http://www.fleets.doe.gov/cgi-bin/fleet/main.cgi?17357,state_ins_rep,5,468050; see also IRS Publication 535.

[18] *Alternative Fuel Vehicle Fleet Buyer's Guide*, http://www.fleets.doe.gov/cgi-bin/fleet/main.cgi?17357,state_ins_rep,5,468050; see also IRS Publication 535.

Exemption of EVs from Luxury Taxes

In addition to the tax incentive provisions under EPAct, the Taxpayer Relief Act of 1997 (P.L. 105-34) amended the Internal Revenue Code to exempt EVs from Federal excise "luxury" taxes and from "luxury" depreciation schedules.[19]

3.1.2 Alternative Fuel Vehicle Requirements for Federal, State, and Alternative Fuel Provider Fleets

EPAct Procurement Requirements for AFVs in Federal Fleets

In addition to the tax provisions established under EPAct, the law mandated Federal, state, and "alternative fuel provider" fleets to purchase AFVs, including EVs. These provisions have been underscored by a series of Executive Orders that further the commitments of Federal agency fleets to adopt AFVs. Likewise, state and alternative fuel provider fleets must meet the requirements outlined in the Alternative Fuel Transportation Program, Final Rule under the EPAct implementing regulations.[20]

Section 303 of EPAct requires Federal agencies to acquire a specified number of AFVs, starting in 1993. Under the Act, the Federal Government was required to acquire at least 5,000 light duty AFVs in FY1993, 7,500 light duty AFVs in FY1994, and 10,000 light duty AFVs in FY1995. Following FY1995, all Federal fleets consisting of 20 or more light duty motor vehicles must meet a specific percentage requirement for AFVs, including: 25 percent in FY1996, 33 percent in FY1997, 50 percent in FY1998, and 75 percent in FY1999 and thereafter.[21] These requirements are summarized in Table 3-3 below. (See "Success of the EPAct AFV Program for Federal Fleets" later in this section for a summary of the success of the EPAct AFV directives.)

Table 3-3. Summary of EPAct Requirements for Federal Government Acquisition of Light Duty AFVs

Fiscal Year Vehicle Acquired	Applicable Fleet	Number of AFVs Required
FY1993	Entire Federal Government	5,000 total, Government-wide
FY1994		7,500 total, Government-wide
FY1995		10,000 total, Government-wide
FY1996	Each Federal fleet with 20 or more light duty vehicles	20% of each fleet
FY1997		33% of each fleet
FY1998		50% of each fleet
FY1999 and thereafter		75% of each fleet

EPAct established a credit system to aid the different agencies in meeting their AFV targets, whereby vehicle procurement agents may receive a specific number of credits when procuring different types of AFVs. Federal and State government agencies that are unable to meet their requirements, as well as alternative fleet providers, may then purchase credits from those agencies or fleet providers that exceed their procurement requirement (see Table 3-5).[22] Furthermore, to encourage and promote the use of AFVs in Federal fleets, EPAct also provides an agency incentive program and a recognition and incentive awards program for Federal agencies. Under the Act, the General

---

[19] See http://www.fourmilab.ch/ustax/www/t26-D-31-A-4001.html.

[20] 10 CFR Part 490.

[21] EPAct §303. Please also see the clean cities web site to for actual numbers of AFVs on the road as of 2002. www.ccities.doe.gov

[22] See EPAct §508.

Services Administration (GSA) may offer a reduction in fees charged to agencies to lease AFVs below those fees charged for the lease of comparable conventionally fueled motor vehicles.[23] The GSA is also required to establish an annual awards program that recognizes Federal employees who have demonstrated "the strongest commitment to the use of alternative fuels and fuel conservation in Federal motor vehicles."[24] Moreover, the Act requires the U.S. Postal Service to provide a report to Congress outlining its AFV program.[25]

Executive Order 13149: Fuel Economy and AFV Procurement Requirements for Federal Fleets

Federal agencies have been required to follow guidelines established in Executive Order 12844 (April 21, 1993) and subsequently reinforced by Executive Order 13031 (December 13, 1996) that underscored the policies and objectives of the Federal agency AFV provisions of EPAct. Both were superceded by Executive Order 13149, signed in April 21, 2000, which further strengthened the Federal government's commitment to promote the use of alternative fuel vehicles in Federal fleets.

Executive Order (E.O.) 13149 requires Federal agencies operating 20 or more motor vehicles within the United States to reduce the fleet's annual petroleum consumption by 20 percent below FY1999 levels by the end of FY2005.[26] To meet this goal, Federal agencies are given significant flexibility in developing an appropriate strategy to meet the petroleum reduction levels. Agencies are required to use alternative fuels, such as electricity, to meet the majority of the fuel requirements for vehicle fleets operating in "metropolitan statistical areas," i.e., metropolitan areas with populations of more than 250,000 in 1995 according to the Census Bureau. Where feasible, the Order also instructs agencies to consider procuring "innovative" alternative fuel vehicles that are capable of large improvements in fuel economy, such as HEVs. Agencies are required to increase the average EPA fuel economy rating of their light-duty vehicle acquisitions by at least one mile per gallon (mpg) by 2002 and 3 mpg by 2005 above 1999 acquisition levels. Agencies are also encouraged to adopt awards and performance evaluation programs that reward federal employees for exceptional performance in implementing the Order.[27] Federal fleet requirements under E.O. 13149 are summarized in Table 3-4.

Table 3-4. Summary of Executive Order 13149 Requirements for Federal Government Fleets		
Applicable Fleet	Effective Date	Action Required
Each Federal fleet with 20 or more light duty vehicles	FY2002	Increase average EPA fuel economy rating of light-duty vehicle acquisitions by 1 mpg above FY1999 levels
Each Federal fleet with 20 or more light duty vehicles	FY2005	Increase average EPA fuel economy rating of light-duty vehicle acquisitions by 3 mpg above FY1999 levels
Each Federal fleet with 20 or more light duty vehicles	By end of FY2005	Reduce fleet's annual petroleum consumption by 20% below FY1999 levels
Each Federal fleet with 20 or more light duty vehicles operating in Metropolitan Statistical Areas	By end of FY2005	Same action as above, but must include alternative fuels to meet majority of fuel requirements

---

[23] EPAct §306.

[24] EPAct §307.

[25] EPAct §311.

[26] E.O. 13149 §201. Independent agencies are encouraged but not required to comply with the Order. §504.

[27] E.O. 13149 §303

E.O. 13149 elaborates on the AFV acquisition credit program with respect to Federal agencies. As established under EPAct (described above), credits received for the acquisition of AFVs (by government or non-governmental entities) are freely transferable among fleet owners and others required to acquire AFVs under the Act.[28] Fleet owners that do not meet the EO acquisition requirements for AFVs may thus purchase credits from fleet owners with a surplus of AFVs credits. Under the Order, agencies receive: (1) one credit for each light-duty AFVs acquired; (2) two credits for each light-duty AFV that exclusively uses an alternative fuel and for each ZEV (see section 3.2 below for a discussion of ZEVs); (3) three credits for dedicated medium-duty AFVs; and (4) four credits for dedicated heavy-duty AFVs.[29] This provision enhances the credit allowances under EPAct, which awards a single credit for each AFV acquired.[30] Table 3-5 summarizes the number of credits available for each type of acquired AFV.

Table 3-5	Summary of Credits for Federal Fleet Acquisitions of AFVs under Executive Order 13149	
Type of AFV	**Number of Credits Awarded**
Each light-duty AFV	1 credit
Each light-duty AFV exclusively using an alternative fuel	2 credits
Each ZEV	2 credits
Each dedicated medium-duty AFV	3 credits
Each dedicated heavy-duty AFV	4 credits

In order to provide for adequate access to refueling infrastructure, Federal agencies are directed under E.O. 13149 to "team with state, local, and private entities to support the expansion and use of" public refueling stations for AFVs.[31] State, local, and private groups may also establish non-public alternative fuel stations if no commercial infrastructure is available in their territory.[32]

Success of the EPAct AFV Program for Federal Fleets

According to the Department of Energy (DOE) Clean Cities Report *Federal Fleet AFV Program Status,* dated June 2, 1998, as of 1998, of more than 570,000 vehicle acquisitions overall, the estimated cumulative total AFV acquisitions in Federal agencies totaled more than 34,000 vehicles between FY1991 and FY1998. This represented about 80 percent compliance with the 44,600 required AFV acquisitions under EPAct. Only several hundred of the AFVs acquired were qualified EVs.[33]

---

[28] EPAct §508.

[29] E.O. 13149 §401.

[30] EPAct §508.

[31] E.O. 13149 §402(a).

[32] E.O. 13149 §402(b).

[33] U.S. Department of Energy, Federal Fleet AFV Program Status (June 2, 1998), available at: http://www.ccities.doe.gov/pdfs/slezak.pdf. As stated in the report:

> Of the 34,000+ AFVs acquired by Federal agencies, approximately 10,000 (30 percent) have been M-85 (methanol mixed with gasoline) flexible fuel vehicles, 6,000 (17 percent) have been E-85 (ethanol mixed with gasoline) flexible fuel vehicles, and 18,000 (52 percent) have been compressed natural gas (CNG) vehicles. Several hundred each of *electric* and liquefied petroleum gas (LPG or propane) vehicles have also been acquired. Projections for future Federal AFV acquisitions, based on discussions with Federal

In January 2002, three environmental organizations filed a lawsuit in Federal court against 17 Federal agencies for allegedly failing to comply with the AFV acquisition requirements imposed under EPAct.[34] The plaintiffs claim that all 17 agencies have failed: (1) to meet their AFV acquisition requirements; (2) to file the necessary compliance reports with Congress; and (3) to make these reports available to the public. The complaint also alleges that DOE failed to complete a required private and municipal AFV fleet rulemaking. As a remedy, the plaintiffs request that the court order the agencies to comply with these requirements, and to require the agencies to offset their future vehicle purchases with the number of AFVs necessary to bring them into compliance with EPAct's acquisition requirements for 1996 through 2001. A decision on the case is pending.[35]

EPAct Procurement Requirements and Incentives for AFVs in Alternative Fuel Provider and State Fleets

In addition to Federal fleet requirements, EPAct established the State and Alternative Fuel Provider (S&FP) Program, a DOE regulatory program that requires covered state and "alternative fuel provider" fleets to purchase AFVs as a portion of their annual light duty vehicle acquisitions.[36] It is important to note, as mentioned above, that HEVs do not qualify as AFVs under the program because they are not *primarily* powered by the electric motor.[37]

As required by EPAct, DOE has developed a mandatory vehicle schedule for acquiring light duty AFVs, including electric vehicles, for alternative fuel providers and states. The mandatory acquisition schedule for *alternative fuel provider* fleets is:

- 30 percent for model year 1997;
- 50 percent for model year 1998;
- 70 percent for model year 1999; and
- 90 percent for model year 2000 and thereafter.[38]

The AFV regulations cover a state agency if it owns or operates 50 or more light-duty vehicles, at least 20 of which are used primarily within a metropolitan area.[39] States are required to prepare plans for implementing an AFV program and various policy incentives that may be used to encourage the adoption of AFVs.[40] The mandatory acquisition schedule of AFVs for *state government* fleets is:

- 10 percent for model year 1997;
- 15 percent for model year 1998;

---

agencies' procurement personnel and manufacturers, indicate that flexible fuel E-85 vehicles will be the most common AFV procured by agencies' to comply with EPACT, followed by CNG. (italics added)
Id.
[34] Center for Biological Diversity v. Abraham, N.D. Cal., No. CV-00027 (January 2, 2002). The agencies named in the suit include: the Departments of Energy, Commerce, Justice, Interior, Veterans Affairs, Agriculture, Transportation, Health and Human Services, Housing and Urban Development, Labor, State, and Treasury; the Environmental Protection Agency; the U.S. Postal Service; the National Aeronautics and Space Administration; the U.S. Nuclear Regulatory Commission; and the General Services Administration.
[35] See www.evaa.org.
[36] EPAct §501; 10 CFR 490.303.
[37] 10 C.F.R. §490.2. See also U.S. Department of Energy Office of Transportation Technologies, *Commercially Available Hybrid Electric, Low-Speed Vehicles not Eligible for EPAct Credit* (September 2002), http://www.nrel.gov/docs/fy01osti/30782.pdf.
[38] 10 CFR 490.302.
[39] see *Federal Register*, Volume 61, Number 51, pages 10627-10628.
[40] EPAct §409.

- 25 percent for model year 1999;
- 50 percent for model year 2000; and
- 75 percent for model year 2001 and thereafter.[41]

Fleets earn credits for each vehicle purchased, and credits earned in excess of their requirements can be banked or traded with other fleets. As with the Federal AFV program, states and alternative fuel providers that exceed EPAct requirements receive additional credits, while those that are unable to meet the requirements by acquiring AFVs may purchase credits from those holding them.[42]

As of FY2002 (MY2001), states and alternative fuel provider fleets have collectively acquired more than 60,000 AFVs since the launch of the program, exceeding the program quota.[43] According to the 2001 Annual Report, only about 9% of the S&FP fleets had failed to comply with program requirements. About 4.5% of AFVs acquired were qualified electric vehicles.[44]

Box 3-1   Calculating the Petroleum Equivalency Factor (PEF)

The PEF methodology was developed by DOE to compare the fuel economy of EVs with that of conventional gasoline vehicles. The PEF equation is:

$$PEF = Eg \ * \ 1 \ / \ 0.15 \ * \ AF \ * \ DPF$$

Where:

Eg   = average fossil fuel electricity generation efficiency * average electricity transmission efficiency * refining and distribution efficiency * watt-hours energy per gallon gasoline conversion factor

= gasoline-equivalent energy content of electricity factor

1/0.15   = Fuel content factor

AF   = Petroleum-based accessory factor

DPF   = Driving pattern factor

3.1.3   Petroleum Equivalency Factors for Electric Vehicles

One significant regulatory development has been the determination of petroleum-equivalent fuel economy values for EVs. These factors can be used by automobile manufacturers in the total calculation of a manufacturer's corporate average fuel economy (CAFE), according to regulations prescribed by EPA and the Department of Transportation. On June 12, 2000, DOE released the final calculation to be used to determine the petroleum-equivalency factor (PEF) for EVs. This procedure is described further in Box 3-1. Under the final PEF calculation, an EV achieving 0.24 kWh/mile and having no petroleum-fueled accessories (e.g., a diesel-fired heater or defroster) would receive a petroleum-based fuel economy value of 335.24 mpg.[45]

---

[41] 10 CFR 490.201.

[42] EPAct §508; 10 CFR 409. See also Alternative Fuel Transportation Program, Final Rule, 10 CFR Part 490) http://www.fleets.doe.gov/cgi-bin/fleet/main.cgi?17357,state_ins_rep,5,468050.

[43] U.S. Department of Energy Office of Transportation Technologies, *What's New: Spring 2002 Update* (May 2002), http://www.ott.doe.gov/epact/pdfs/whatsnew_spring_02.pdf.

[44] U.S. Department of Energy Office of Transportation Technologies, Program Activity and Accomplishments in FY2001 (December 2001), http://www.ott.doe.gov/epact/pdfs/fy01rpt.pdf.

[45] 36986 Federal Register / Vol. 65, No. 113 / Monday, June 12, 2000; available at http://www.ott.doe.gov/legislation.shtml

3.1.4 Pending Federal Legislation and Programs

Several pieces of legislation have been introduced by the 107th Congress that would enhance existing legislation affecting EV and HEV use in the U.S., including several that specifically promote the use of battery-powered EVs (BEVs), fuel cell vehicles, and HEVs. The most prominent of these are the Securing America's Future Energy Act of 2001 (HR 4), and the Advanced Motor Vehicle Technology and Alternative Fuels Consumer Incentives Act (S. 760). Both would extend the Federal tax credit of $4,000 for the purchase of light-duty EVs to 2007, with HR 4 providing an additional $1,000 tax credit for EVs with a driving range of 70 miles or higher for a single charge, and S. 760 providing an additional $2,000 tax credit for EVs with a driving range of 100 miles or higher for a single charge. Higher tax credits are available for heavier-duty BEVs. S. 760 would also provide a 10 percent tax credit of up to $4,000 for purchase of neighborhood EVs. Base tax credits of $4,000 are available for fuel cell vehicles in both bills with up to an additional $4,000 depending on the fuel economy increase over conventional vehicles. The credits for fuel cell vehicles are dependent on meeting Tier 2, Bin 5 emission standards. Tax credits for fuel cell vehicles under both bills would extend through 2011. Tax credits for HEVs in S. 760 are dependent on the power of the electric drive portion of the powertrain, and the increase in fuel economy relative to conventional vehicles, and on meeting Tier 2, Bin 5 emission standards starting in 2004. A maximum of $4,000 in tax credits would be available for HEVs in S. 760. HR 4 has very similar provisions except that it does not have an emissions requirement, and it adds a "conservation credit" of up to $500 dependent on the lifetime fuel savings of the HEV. In total, an HEV could get up to $5,000 under HR 4. HR 4 would also extend tax deductions available for development of clean fuel infrastructure through 2007.

3.2 State Policies and Programs

3.2.1 California

Low Emission Vehicle (LEV) and Zero Emission Vehicle (ZEV) Regulations

With the exception of the State of California, Section 209(a) of the Federal Clean Air Act (CAA) prohibits states from adopting or enforcing standards for new motor vehicles or new motor vehicle engines.[46] In response to California's severe air pollution problems, CAA Section 209(b) grants the State the explicit authority to set its own standards for vehicular emissions, so long as the standards are equal to, or more stringent than, those set by the CAA and are approved by EPA.[47] State studies have found that about half of smog-forming pollutants are produced by gasoline and diesel-powered vehicles, and that only alternative technologies would help California reduce motor vehicle air pollution that will result from increasing driving rates in the State.[48]

As provided in the Clean Air Act, other states are permitted to follow California so long as any motor vehicle emissions regulations adopted by those states are identical to California's.[49] Since California introduced its LEV standards in 1990, four other States— New York, Massachusetts, Maine, and Vermont—have adopted the California emissions requirements for a percentage of motor vehicles sold in those states (see Section 3.2.2).

---

[46] 42 U.S.C. 7609(a).
[47] 42 U.S.C. 7609(b).
[48] See California's Zero Emission Vehicle Program, CARB, Fact Sheet, 12/06/01 http://www.arb.ca.gov/msprog/zevprog/factsheet/evfacts.pdf.
[49] 42 U.S.C. 7507.

LEV I Regulatory Program

The flexibility provided to California under the CAA paved the way for sweeping regulation that has established extensive standards for low and zero emissions vehicles sold in the State. Under CAA authority, in 1990 the California Air Resources Board (CARB) adopted regulations to require automobile manufacturers to introduce low-emission vehicles (LEVs)—and ZEVs—to the automobile market in the State. The regulations would require manufacturers to sell a certain percentage of these vehicles each year. Known as LEV I, the new standards promised to introduce EVs, HEVs, and various other low emission vehicles, and to affect the entire automobile market in California.

LEV I standards are based on the introduction of four classes of vehicles with increasingly more stringent emissions requirements. These include:

- transitional low emissions vehicles (TLEVs);
- low-emission vehicles (LEVs);
- ultra-low-emission vehicles (ULEVs); and
- zero emissions vehicles (ZEVs)[50].

Under the LEV I requirements, as of 1994 manufacturers are permitted to certify vehicles in any combination of the LEV categories through 2003 in order to satisfy the LEV standard.[51] It should be noted that under current regulations, auto manufacturers are also required to comply with a fleet-based average Non-Methane Organic Gas standard (NMOG), which introduces more and more stringent standards with each model year.[52]

LEV II Regulatory Program

Following a hearing in November 1998, the CARB amended the LEV I regulations and adopted LEV II, the second-generation LEV program. While the first set of LEV standards covered 1994 through 2003 models years, the LEV II regulations cover 2004 through 2010 and represent continued emissions reductions. The LEV II amendments were formally adopted by the CARB on August 5, 1999 and came into effect on November 27, 1999.[53]

The more stringent LEV II regulations were adopted in part to keep up with changing passenger vehicle fleets in the state, where more sport utility vehicles (SUVs) and pickup trucks are used as passenger cars rather than work vehicles. The LEV II standards were a necessary step for the state to meet the Federally-mandated CAA goals that address ambient air quality standards as outlined in the 1994 State Implementation Plan (SIP).[54] LEV II increased the stringency of the emission standards for all light- and medium-duty

---

[50] EVs provide the only automobile technologies available today that can meet the ZEV standard. See the California Health and Safety Code, Sections 39656-39659.

[51] See California Air Resources Board, *California Exhaust Emissions Standards and Test Procedures for 2001 and Subsequent Model Passenger Cars, Light-Duty Trucks, and Medium-Duty Vehicles*, Proposed Amendments (Sept. 28, 2001).

[52] §1960.1(g)(2). California's fleet average NMOG mechanism "requires manufacturers to introduce an incrementally cleaner mix of Tier 1, TLEV, LEV, ULEV and ZEV vehicles each year, with the fleet average NMOG value for passenger cars and lighter light-duty trucks decreasing from 0.25 gram/mile in the 1994 model year to 0.062 gram/mile in the 2003 model year." See California Air Resources Board, *The California Low-Emission Vehicle Regulations* (May 30, 2001), http://www.arb.ca.gov/msprog/levprog/cleandoc/levregs053001.pdf.

[53] Low-Emission Vehicle Program website (September 28, 2001), located at http://www.arb.ca.gov/msprog/levprog/levprog.htm.

[54] Low-Emission Vehicle Program website (September 28, 2001), located at http://www.arb.ca.gov/msprog/levprog/levprog.htm.

vehicles beginning with the 2004 model year and expanded the category of light-duty trucks up to 8,500 lbs. gross vehicle weight (including almost all SUVs) to be subject to the same standards as passenger cars.[55] When LEV II is fully implemented in 2010, it is estimated that smog-forming emissions in the Los Angeles area will be reduced by 57 tons per day, while the statewide reduction is expected to be 155 tons per day.[56]

The LEV II standards go further to require that vehicles classified as LEV and ULEV meet NO_x standards which are 75 percent below LEV I requirements based on fleet averages. In addition, fleet average durability standards are extended from 100,000 to 120,000 miles. LEV II also allows manufacturers to receive credits for vehicles meeting near-zero emissions, such as fuel cell HEVs, and a new category of vehicles called super ultra-low emissions vehicles (SULEVs).[57] The LEV II standards were also designed to respond to some delays and "inertia" the LEV program had been facing, and pushed back the starting year of the program to 2003.

Under LEV II, manufacturers may certify vehicles under one of five emission standards, listed in order from least to most stringent:

- transitional low emissions vehicles (TLEVs)
- low-emission vehicles (LEVs);
- ultra-low-emission vehicles (ULEVs);
- super ultra-low emissions vehicles (SULEVs); and
- zero emissions vehicles (ZEVs).

Some examples of LEV I and LEV II emissions standards for the different vehicles types are provided in Tables 3-6 and 3-7.

---

[55] California Air Resources Board: Notice Of Public Hearing To Consider The Adoption Of Amendments To The Low-Emission Vehicle Regulations, November 15, 2001. The California Low-Emission Vehicle Regulations.
(As of May 30, 2001) (available at: http://www.arb.ca.gov/msprog/levprog/test_proc.htm)

[56] LEV Program, http://www.arb.ca.gov/msprog/levprog/levprog.htm. See also The California Low-Emission Vehicle Regulations, (As of May 30, 2001) (available at: http://www.arb.ca.gov/msprog/levprog/test_proc.htm)

[57] See California Air Resources Board, *California Exhaust Emissions Standards and Test Procedures for 2001 and Subsequent Model Passenger Cars, Light-Duty Trucks, and Medium-Duty Vehicles*, Proposed Amendments (Sept. 28, 2001).

Table 3-6.	LEV I Exhaust Emission Standards for New MY2001–MY2003 Passenger Cars and Light Duty Trucks (3,750 lbs. LVW or less)					
Durability of Vehicle	Vehicle Emission Category	NMOG (g/mi)	Carbon Monoxide (g/mi)	NOx (g/mi)	Formaldehyde (mg/mi)	Particulates fr. diesel vehicles (g/mi)
50,000	Tier 1	0.250	3.4	0.4	n/a	0.08
	TLEV	0.125	3.4	0.4	15	n/a
	LEV	0.075	3.4	0.2	15	n/a
	ULEV	0.040	1.7	0.2	8	n/a
100,000	Tier 1	0.310	4.2	0.6	n/a	n/a
	Tier 1 diesel option	0.310	4.2	1.0	n/a	n/a
	TLEV	0.156	4.2	0.6	18	0.08
	LEV	0.090	4.2	0.3	18	0.08
	ULEV	0.055	2.1	0.3	11	0.04

In order to meet these standards, several car manufacturers developing HEVs and ZEVs have already begun to market their products as one of the categories listed above. As of the present time, the Toyota Prius HEV fully meets SULEV standards in California and exceeds ULEV requirements by about 75 percent.[58] The Honda Insight HEV meets the ULEV standards.[59]

Table 3-7	LEV II Exhaust Emission Standards for New MY2001–MY2003 Passenger Cars and Light Duty Trucks (8,500 lbs. GVW or less)					
Durability of Vehicle	Vehicle Emission Category	NMOG (g/mi)	Carbon Monoxide (g/mi)	NOx (g/mi)	Formaldehyde (mg/mi)	Particulates fr. diesel vehicles (g/mi)
50,000	LEV	0.075	3.4	0.05	15	n/a
	LEV Option 1	0.075	3.4	0.07	15	n/a
	ULEV	0.040	1.7	0.05	8	n/a
120,000	LEV	0.090	4.2	0.07	18	0.01
	LEV Option 1	0.090	4.2	0.10	18	0.01
	ULEV	0.055	2.1	0.07	11	0.01
	SULEV	0.010	1.0	0.02	4	0.01
150,000 (optional)	LEV	0.090	4.2	0.07	18	0.01
	LEV Option 1	0.090	4.2	0.10	18	0.01
	ULEV	0.055	2.1	0.07	11	0.01
	SULEV	0.010	1.0	0.02	4	0.01
	LEV	0.090	4.2	0.3	18	0.08
	ULEV	0.055	2.1	0.3	11	0.04

---

[58] See http://www.toyota.com/.
[59] See http://www.hondacars.com/.

Zero Emission Vehicle Mandate

Possibly the most controversial element of the LEV program has been the Zero Emission Vehicle requirement, which began with LEV I and was amended for LEV II. This requirement, known as the "ZEV Mandate" requires that a specific minimum percentage of passenger cars and the lightest light-duty trucks marketed in California by large or intermediate volume manufacturers be ZEVs. (As noted below, requirements differ based on the manufacturer's volume of sales.)[60] To initiate the process for car manufacturers to begin to adapt to the new ZEV requirements, the program at first required car manufacturers to implement a number of small demonstration fleets of ZEVs in the early 1990s and then to gradually implement efforts to market ZEVs to the general public starting in 2003. With the adoption of the newer LEV II regulations, ZEVs considered in the program now include:

- Pure ZEVs (ZEVs)—vehicles with no tailpipe emissions whatsoever;
- Partial ZEVs (PZEVs)—vehicles that qualify for a partial ZEV allowance of at least 0.2 (before an additional "early introduction phase-in multiplier" or "high-efficiency multiplier" are applied to the allowance); and
- Advanced Technology PZEVs (AT PZEVs)—any PZEV with an allowance greater than 0.2.[61]

Pure ZEVs must produce zero exhaust emissions of any criteria or precursor pollutant under any and all possible operational modes and conditions. AT PZEVs include compressed natural gas, HEVs, and methanol fuel cell vehicles. In order to qualify as a PZEV, the AT PZEVs would also have to meet the SULEV tailpipe emissions standard, achieve zero evaporative emissions and include a 150,000-mile warranty for emission control equipment.[62] The Executive Officer of CARB is responsible for certifying new 2003 and all subsequent model year (MY) ZEVs.[63]

The total required volume of a manufacturer's production and delivery for sale of Passenger Cars (PCs) and Light-Duty Trucks 1 (LTD1s) is based on the average from the previous three-year period. The production average is used only for the ZEV requirement. The manufacturer may also choose an alternative to the three-year averaging approach by choosing to base production volumes on an annual basis, using the first year in the three year period and every year thereafter, respectively.

The original LEV I regulations required that specific percentages of all PCs and LDT1s, MY1998 and later, be certified as ZEVs. Under the original rulemaking, the required percentages were: 2 percent of the total volume of a manufacturer's production and delivery for sale for 1998-2000 model year vehicles, 5 percent of the total volume for 2001-2002 model year vehicles, and 10 percent of the total volume for 2003 and subsequent model year vehicles.

---

[60] See California Air Resources Board, *Notice Of Public Hearing To Consider The Adoption Of Amendments To The Low-Emission Vehicle Regulations*, November 15, 2001.

[61] California Air Resources Board, *California Exhaust Emission Standards and Test Procedures for 2003 and Subsequent Model Zero-Emission Vehicles, and 2001 and Subsequent Model Hybrid Electric Vehicles, in the Passenger Car, Light-Duty Truck, and Medium-Duty Vehicle Classes.* (Amended: April 12, 2002), pages A,B-1 to A,B-2. (hereinafter *California Exhaust Emission Standards and Test Procedures*). Qualified PZEVs meet SULEV, evaporative emissions, and on-board diagnostic standards, and offer an extended warranty of 15 years or 150,000 miles, whichever occurs first. See Id., page C-4.

[62] Zero Emission Vehicle Program Changes; ARB, Fact Sheet, 12/10/01 http://www.arb.ca.gov/msprog/zevprog/factsheet/zevchanges.pdf. Note, the current Toyota Prius and Insight HEV models do not yet meet all of the requirements needed to earn either PZEV or AT-PZEV credits. Id.

[63] *California Exhaust Emission Standards and Test Procedures*, page C-1.

Table 3-8	Comparison of Percentage Requirements for Certified ZEVs under LEV I and LEV II[64]	
Model Years	**Original** **LEV I Percentage Requirement**	**Current** **LEV II Percentage Requirement**
1998-2000	2%	Eliminated
2001-2002	5%	Eliminated
2003-2008	10%	10%
2009-2011	10%	11%
2012-2014	10%	12%
2015-2017	10%	14%
2018 and subsequent years	10%	16%

In a 1996 rulemaking, the CARB eliminated the 2 percent and 5 percent requirements for the 1998-2002 model years due to the unlikelihood of compliance, but still maintained the 10 percent requirements for the 2003 and subsequent model years.[65] Between 1998 and 2001, the CARB approved several amendments to the original ZEV regulations that would take form under LEV II. These amendments significantly *reduce* the number of full function ZEVs that will be required in the initial years of the program, but nevertheless institute a gradual *increase* in the minimum required percentage of ZEVs in sales fleets— from 10 percent in 2003 up to 16 percent in 2018.[66] As of Summer 2002, these most recent June 1, 2001 amendments are still pending, but are expected to be adopted without significant additional changes.[67] LEV II requirements are compared with the LEV I requirements in Table 3-6.

Unlike the previous regulations, the most recent amendments require large and intermediate volume manufacturers to meet different percentage of sales requirements for pure ZEVs, PZEVs, and AT PZEVs.[68] Under the latest proposals, major automakers (those selling 35,000 or more passenger cars and light-duty trucks annually in California) could meet the 10 percent requirement for ZEVs sold in the State by selling 20% of their ZEV vehicles as pure ZEVs, 60% as PZEVs, and 20% as AT PZEVs. Intermediate automakers (those selling 4,501 to 35,000 passenger cars and light-duty trucks annually in California) could meet their entire ZEV requirement with PZEV credits, and manufacturers selling fewer than 4,500 vehicles annually would not have to meet any ZEV requirement.[69] Table 3-8 summarizes these requirements. (Small and independent low volume manufacturers are exempt from the ZEV requirements but can acquire credits for the sale of ZEVs or PZEVs).

[64] *California Exhaust Emission Standards and Test* Procedures, page C-1. See also Zero Emission Vehicle Program Changes; ARB, Fact Sheet, 12/10/01.
http://www.arb.ca.gov/msprog/zevprog/factsheet/zevchanges.pdf.
[65] The California Low-Emission Vehicle Regulations (as of May 30, 2001), available at: http://www.arb.ca.gov/msprog/levprog/test_proc.htm.
[66] The California Low-Emission Vehicle Regulations, (as of May 30, 2001), available at: http://www.arb.ca.gov/msprog/levprog/test_proc.htm.
[67] Telephone interview with Tom Evashenk, Staff, CARB (March 5, 2002).
[68] *California Exhaust Emission Standards and Test Procedures*, page C-2.
[69] SB 1782 (1998), see http://www.fleets.doe.gov/fleet_tool.cgi?$$,benefits,

Table 3-9   Summary of ZEV Requirements under LEV II[70]

Applicable Manufacturer	Model Year	Percentage of Sales Required for Compliance
Large Volume Manufacturers	2003-2008	20% of sales as ZEVs (or ZEV credits) at least 20% of sales in additional ZEVs or AT ZEVs (or credits for such vehicles) remaining percentage (up to 60%) of sales as PZEVs (or PZEV credits)
Intermediate Volume Manufacturers	2003 and afterwards	up to 100% PZEV allowance vehicles (or credits)
Small Volume and Independent Low Volume Manufacturers	No requirements, but can acquire credits for sale of ZEVs or PZEVs	

The 2001 amendments also added the category of Light Duty Truck 2 (LDT2) to the original PC and LTD1 categories of vehicles.[71] As a result of the LDT1 and LDT2 categories, all sizes of SUVs and mini-vans would be covered by the LEV II regulations. LDT2 vehicles will be phased in gradually, starting with 17 percent in 2007 and reaching total incorporation by 2012.[72]

Table 3-10   Percentage of LDT2s Required to be Phased in, by model year

2007	2008	2009	2010	2011	2012
17%	34%	51%	68%	85%	100%

The newly proposed regulations would also push back the start date for several requirements, such as the number of PZEV vehicles required in the early years. PZEVs can now be phased in at 25 percent of the previously required level in 2003, and 50 percent, 75 percent, and 100 percent of the previous level in 2004, 2005, and 2006, respectively. Beginning in 2007, automobile manufacturers must also include heavier SUVs, pickup trucks, and vans in the sales figures used to calculate each automaker's ZEV requirement. In other words, in order to sell more SUVs and other heavier vehicles, each automaker must also sell more ZEVs.[73]

Finally, in order to ensure effective cooperation between the State of California and auto manufacturers in implementing the LEV regulatory program, and to encourage continued research and development, demonstration, and commercialization of low and zero emission vehicle technologies, the State entered a separate memorandum of agreement (MOA) with each of the seven largest auto manufacturers. Each MOA represents a commitment between the auto manufacturer and the CARB to ensure the successful

---

[70] *California Exhaust Emission Standards and Test Procedures*, page C-2.

[71] Under California regulations, LTD1 vehicles include any light duty truck up to 3,750 lbs. loaded vehicle weight. LTD2 is defined as any light-duty truck above 3,750 lbs. loaded vehicle weight. U.S. Environmental Protection Agency, Office of Transportation and Air Quality, *Exhaust and Evaporative Emission Standards*, EPA420-B-00-001 (February 2000), located at: http://www.epa.gov/otaq/cert/veh-cert/b00001i.pdf.

[72] *California Exhaust Emission Standards and Test Procedures*, page C-2.

[73] Zero Emission Vehicle Program Changes; CARB, Fact Sheet, 12/10/01 http://www.arb.ca.gov/msprog/zevprog/factsheet/zevchanges.pdf.

launch and long-term success of the ZEV program. These auto manufacturers are Chrysler, Ford, General Motors, Honda, Mazda, Nissan and Toyota.[74]

ZEV Compliance

Auto manufacturers are subject to civil penalties of $5,000 for each sale, attempt of sale, or offer of sale of vehicles failing to meet applicable emissions standards.[75]

ZEV Incentive Programs

Credits. Like the Federal Alternative Fuel Vehicle program, the California program includes a range of credits that provide incentives for the development of ZEV vehicles with improved range and refueling capacity. The amended California ZEV program envisions awarding additional credits for ZEVs introduced ahead of schedule. Automakers will receive four times the normal number of credits for each ZEV introduced in 2001-2002, and 1.25 times the normal number of credits for each ZEV introduced between 2003 and 2005. The provisions also reduce the minimum number of extra credits available for ZEV models with extended ranges of 50 or more miles to 100 or more miles, and provide 10 credits for ZEVs with ranges of 275 or more miles. Extra credits are also awarded for vehicles that can refuel or charge in less than 10 minutes for a 60-mile range. Credits available for small, neighborhood EVs (NEVs) with limited speed and range are increased from one credit per vehicle to: 4.0 credits for each NEV introduced in 2001-2002; 1.25 credits in 2003; and 0.625 credit for 2004-2005; and 0.15 credit thereafter. ZEVs that remain on the road in California for more than three years also receive additional credits.[76]

Grants. The CARB recently took steps to complement recent regulatory amendments to the LEV II program with financial incentives that would encourage consumers to purchase ZEVs prior to the mandated start year of 2003. The CARB is setting up a $38 million program to provide incentives to consumers who are interested in buying or leasing ZEVs. This would add to the $20 million in the Governor's 2001-2002 budget and $18 million already planned for incentives. To help consumers defray the cost of some types of ZEVs, the incentive programs will provide grants of up to $9,000 over three years for ZEVs leased prior to 2003. Grants of up to $5,000 would be available thereafter.[77] A significant number of State and local government grant programs provide additional financial incentives to consumers for the purchase of ZEVs.[78]

Carpool Lanes. An added incentive for the use of ZEVs, ULEVs, and SULEVs was the recent adoption of a law in California that allows single-occupant use of High Occupancy Vehicle (HOVs) lanes by certain electric and AFVs. Use of these lanes normally requires that vehicles have at least two occupants. In order to use these lanes with only one occupant, eligible vehicle owners must obtain an identification sticker from the California Department of Motor Vehicles. Although HEVs such as the Toyota Prius and Honda Insight do not qualify for the special use of HOV lanes, over 55 ZEVs, ULEVs, SULEVs, and compressed natural gas vehicle models do.[79]

---

[74] http://www.arb.ca.gov/msprog/zevprog/factsheet/moa.htm.
[75] California Health & Safety Code, §43211.
[76] Zero Emission Vehicle Program Changes; ARB, Fact Sheet, 12/10/01 http://www.arb.ca.gov/msprog/zevprog/factsheet/zevchanges.pdf.
[77] Zero Emission Vehicle Program Changes; ARB, Fact Sheet, 12/10/01, http://www.arb.ca.gov/msprog/zevprog/factsheet/zevchanges.pdf.
[78] see ARB, Local, State and Federal Zero-Emission Vehicle Incentives http://www.arb.ca.gov/msprog/zevprog/incentiv.htm.
[79] California Air Resources Board, AB71 Single Driver Sticker, Qualifying Vehicles for Carpool Lane use web page, at http://www.arb.ca.gov/msprog/carpool/carpool.htm.

Installation of EV Recharging Infrastructure

In 1994 the California Energy Commission (CEC) became aware of problems with installing EV infrastructure while implementing an early EV demonstration program. Without explicit direction in the California Building Standards governing the proper installation of electric vehicle charging and supply equipment (California Code of Regulation, Title 24), there were inconsistent requirements imposed by building departments from different jurisdictions that oversee electricity usage and EV charging infrastructure.[80] As a result of these concerns, the CARB has recently adopted a series of rules to standardize and create incentives for the development of EV infrastructure. Regulations going into effect in 2006 require "on-board conductive charging" as the standardized charging system for EVs in California. ZEVs qualifying for one or more credits and all grid-connected HEVs (referred to as extended range HEVs in California regulations) will need to be equipped with a conductive connector vehicle inlet.[81] A number of demonstration programs are currently being implemented in the State to identify opportunities for effective EV infrastructure development.[82]

Regulation of Greenhouse Gas Emissions from Motor Vehicles

On July 11, 2002, the California Legislature passed landmark legislation to propose adopting the first GHG emission regulations on motor vehicles in the United States. AB 1493, expected to be signed into law by the Governor of California at the time of publication of this report, could significantly enhance the objectives of the State's LEV and ZEV program. The law requires the CARB to adopt regulations for carbon dioxide emissions from passenger cars, light trucks, and SUVs by January 1, 2005. The bill directs the CARB to adopt regulations "that achieve the maximum feasible reduction of GHGs emitted by passenger vehicles and light-duty trucks and any other vehicles" [83] in the state. The law would take effect January 1, 2006 and would apply to vehicles manufactured in the 2009 model year and after. One interesting condition in the final legislation is to require CARB to develop regulations that specifically do not: (1) impose additional fees or taxes on motor vehicles, fuel, or miles traveled; (2) ban the sale of any vehicle category in the state; (3) require reductions in vehicle weight; (4) limit speed limits; or (5) limit vehicle miles traveled. AB 1493 would also require the California Climate Action Registry to develop procedures by July 1, 2003, in consultation with CARB, for the reporting and registering of vehicular GHG reductions to the Registry. (The California Registry is described in greater detail in Section 3.3) As stipulated in the Clean Air Act, once AB 1493 is signed into law, other states would be able to follow California in adopting equally stringent regulation of carbon dioxide emissions from automobiles.

3.2.2 Adoption of California LEV II Standards in Northeastern States

As discussed, California is the only State with the ability to adopt motor vehicle emissions standards that exceed those of the CAA.[84] However, under Section 177 of the CAA other States are permitted to adopt any regulations to address motor vehicle emissions that are enacted and adopted by California, so long as the regulations are no more stringent than California's standards and a two-year lead-time is provided prior to the date the regulations come into effect. In the early 1990s, New York, Massachusetts, Maine, and Vermont adopted the California LEV standards.

---

[80] http://www.afdc.doe.gov/altfuel/ele_standard.html.
[81] Electric Vehicle Association of the Americas, www.evaa.org.
[82] U.S. Department of Energy, http://www.fleets.doe.gov/fleet_tool.cgi?$$,benefits,1
[83] California, AB 1058 (as amended, May 31, 2001).
[84] 42 U.S.C. 4709(b).

3 Regulatory and Policy Frameworks

With the exception of Maine, which has repealed its California-based ZEV regulations,[85] each of those states has adopted the 10 percent ZEV sales mandate commencing in model year 2005, two years after the California start year of 2003. In 2000 and 2001, respectively, New York and Massachusetts took the further steps of adopting California's LEV II regulations, as amended.[86] Vermont has yet to adopt the most recently amended LEV II regulations, but is expected to do so in 2002. Beginning in model year 2005, New York also will require the LEV II program for medium-duty vehicles, including larger pick-up trucks and SUVs weighing between 8,500 and 14,000 pounds.[87]

To date, New York and Massachusetts have adopted regulations that would provide automobile manufacturers greater flexibility in complying with the ZEV mandate. Manufacturers can choose to comply with either the California ZEV mandate beginning in model year 2005, or can opt into what is called the northeast states' ZEV Alternative Compliance Plan (ACP) in model year 2004, as explained in Table 3-9 below. In either case, manufacturers will be required to implement the full California ZEV mandate in model year 2007.[88]

Table 3-11	Summary of Alternative Compliance Plan for ZEVs in New York and Massachusetts[89]	
Model Year	Type of Vehicle	Percentage Requirements
2004	PZEVs	10% of all vehicle sales
2005	PZEVs	9% of all vehicle sales
	AT PZEVs or pure ZEVs	1% of all vehicle sales
2006	PZEVs	7% of all vehicle sales
	AT PZEVs	2% of all vehicle sales
	pure ZEVs	1% of all vehicle sales
2007	PZEVs	6% of all vehicle sales
	AT PZEVs	2% of all vehicle sales
	pure ZEVs	2% of all vehicle sales

Any manufacturer opting to use the ACP will be required to submit a projected compliance report at the beginning of each model year. The ACP option also allows manufacturers to meet up to 25 percent of their ZEV requirements with Infrastructure and Transportation System Projects that place advanced technology vehicles in service.

---

[85] See State of Maine Department of Environmental Protection, Rule Chapter 127, *New Motor Vehicle Emission Standard,* Basis Statement for Amendments of December 21, 2000.

[86] In 1993, Maryland and New Jersey also adopted the California LEV program, provided that surrounding States also adopt the California standards. EVAA, State Laws and Regulations Impacting Electric Vehicles (January 2002), http://www.evaa.org.

[87] Governor: Regulation to Reduce Harmful Vehicle Emissions, Alternative to Promote Clean Vehicle Technology, Improve Air Quality (January 4, 2002), http://www.state.ny.us/governor/press/year02/jan4_02.htm; See also New York Adopts New California Emission Standards, EarthVision Environmental News, November 29, 2000, http://www.climateark.org/articles/2000/4th/nyadnewc.htm.

[88] See Background Document and Technical Support For: Public Hearings on the Amendments to the State Implementation Plan for Ozone; and Hearing and Findings under the Massachusetts Low Emission Vehicle Statute - 310 CMR 7.40: The Massachusetts Low Emission Vehicle Program (February 2002), http://www.state.ma.us/dep/bwp/daqc/daqcpubs.htm.

[89] Governor: Regulation to Reduce Harmful Vehicle Emissions, Alternative to Promote Clean Vehicle Technology, Improve Air Quality (January 4, 2002), http://www.state.ny.us/governor/press/year02/jan4_02.htm.

3.2.3   Other State Programs

Over 25 states throughout the U.S. and the District of Columbia have adopted regulations that encourage the use of EVs and HEVs. More than ten states now have laws in place that provide tax incentives to individuals or businesses for the purchase of AFVs, including EVs. In addition to the California LEV II regulations, typical state regulations include the following (note, state abbreviations are provided for each applicable regulation below): [90]

- access to HOV lanes for HEVs, EVs, and other LEVs at any time, and regardless of the number of people occupying the vehicle (AZ, GA, UT, VA);
- exemptions for EVs from parking and other fees (HI);
- individual or business tax incentives, including tax credits or deductions, for the purchase of AFVs and LEVs, including EVs (AZ, GA, KS, LA, ME, MD, NY, OK, OR, UT, VA);
- individual or business tax incentives, including tax credits or deductions, for the construction of AFV and LEV fuel delivery systems (AZ, LA, RI, VA);
- tax incentives, including tax credits or deductions, for manufacturers of AFVs and LEVs (AK, MI);
- tax credits for each job created in manufacturing clean fuel vehicles or converting vehicles to operate on clean fuels (VA);
- exemption of state and/or local sales tax for the purchase of AFVs or AFV conversion equipment (AZ, NH, PA);
- adjustments to fuel taxes to reflect use of AFVs (HI);
- grants to businesses, individuals, local governments, and non-profit organizations towards the purchase of AFVs or AFV fleets (AZ, CA, PA);
- regulations to facilitate the commercialization of AFVs, including EVs and HEVs (NH);
- requirements for state and municipal fleets to acquire AFVs and LEVs, to convert fleets to AFVs, to meet specific clean fuel standards, or to develop AFV infrastructure (DC, LA, MA, MI, MO, NV, NH, NM, NY, OK);
- exemption for certain AFVs or LEVs from emissions inspections and other motor vehicle registration fees and requirements (AZ);
- regulations addressing clean fuel vehicle identification labels or decals (CA);
- exemptions from vehicle registration requirements for AFVs, including EVs, neighborhood electric vehicles (NEVs), and electric motor golf carts (AZ);
- permission for certain types of AFVs, such as electric scooters, to be ridden on public streets (CA);
- special requirements for public utilities to adopt and/or promote LEVs (CA); and
- research programs for the study of AFV technologies (SC, TN).

3.3   Relevant Domestic and International Climate Change Policy and Market Developments

U.S. and international climate change policy could have a dramatic influence on the development of EVs and HEVs and their expanded use throughout the world. Significant international attention has been given to the adoption of the Kyoto Protocol—the binding international framework for implementing specific actions to reduce a country's GHG emissions—by parties to the UNFCCC. Excluding the U.S., most parties to the UNFCCC have either ratified or have expressed commitments to ratify the Kyoto Protocol. However, in 2001, President Bush announced complete U.S. withdrawal from the Kyoto Protocol, instead offering that the U.S. would develop an alternative approach to reducing

---

[90] For a full list of States with related laws and regulations, and a description of each, see EVAA, *State Laws and Regulations Impacting Electric Vehicles* (January 2002), http://www.evaa.org.

domestic GHG emissions. At the time of publication of this report, many of the details of that domestic policy are still under development, but generally include the following key components:

- a commitment to reduce GHG emissions intensity—the ratio of GHG emissions to economic output—by 18 percent over ten years;
- improvements to the U.S. national GHG emissions registry (reporting) program, known as the Voluntary Reporting of GHGs "1605(b)" Program (established under Section 1605(b) of EPAct), now implemented by the Energy Information Administration (EIA) in DOE;
- protection and provision of transferable credits for GHG emission reductions under a future climate change regime; and
- a commitment of financial and technical resources for the continued research of climate change and innovative new technologies to reduce GHG emissions.[91]

In addition to policy developments at the national level, a number of U.S. States and local communities have introduced various legislative initiatives to reduce GHG emissions. These initiatives often reach even further than the measures proposed by the Bush Administration. Most of the policies implemented to date, target the electricity sector by capping emissions from power plants (Massachusetts and New Hampshire) or by setting emissions standards for new facilities (Oregon).[92] However, as described in Section 3.2.1, the State of California recently passed a bill to reduce GHG emissions from the transportation sector, mandating improved efficiency standards. This bill is the first policy initiative in the U.S. to directly influence GHG emissions from the transportation sector. As such, it is likely to have a significant impact on the direction of GHG policy initiatives in the rest of the country and on the future adoption of EVs and HEVs.

With respect to GHG registries, a number of efforts are already under way that may contribute to the development of the national GHG emissions reporting program suggested by the President. In addition to the 1605(b) Program, the State of California recently established the California Climate Action Registry, an independent, non-profit organization dedicated to working with industries, power generators, governmental bodies, and others operating in the State of California to develop a systematic and effective GHG emissions reporting system. The newly formed reporting system is designed to include reporting for GHG emissions reductions from clean transportation, such as the adoption of EVs and HEVs in motor vehicle fleets, in addition to industrial combustion activities and electricity consumption. Various other State GHG emissions registries have also been proposed, as well as an alternate Federal registry under the new EPA Climate Leaders Program.[93]

---

[91] President Announces Clear Skies & Global Climate Change Initiatives, National Oceanic and Atmospheric Administration (Silver Spring, Maryland, February 14, 2002), available at: http://www.whitehouse.gov/news/releases/2002/02/20020214-5.html.

[92] On January 2, 2002, the New Hampshire House passed HB 284—the New Hampshire Clean Power Act. This multi-pollutant legislation affects six units at three different facilities and includes the following emission reduction requirements by 2006: a 75% reduction in SO_2 (3.0 lb/MWh); a 70% reduction in NO_X (1.5 lb/MWh); and, a 3% reduction in CO_2 compared to 1990 levels. In April, 2001, Massachusetts passed a bill requiring the state's 6 highest emitting power plants to: 1) cap their CO_2 emissions at historical levels, and 2) lower their emissions rate to 1,800 lb CO_2/MWh by 2006 or 2008, with a goal of reducing total plant emissions by 10%. In 1997, the State of Oregon required all new power plants to comply with a CO_2 emission standard, specifying that all new natural gas facilities must have an emissions rate that is at least 17% below the most efficient base load gas plant operating in the U.S. Power plants in all three states are allowed to purchase emission offsets from third-party entities to satisfy these requirements.

[93] Id.

With regard to federal support for technology research and development, increased funding levels could potentially lead to advances in HEV, EV, fuel cell, and related clean fuel technologies that help reduce GHG emissions from motor vehicles. The President's budget in FY2003 provides for $4.5 billion for global climate change-related activities, a $700 million increase over previous years.[94]

It is important to note that, in addition to these recent domestic policy activities, increased international activity to implement the Kyoto Protocol could be a potentially important driver for increased development and implementation of EVs and HEVs in overseas markets. This, in turn, could have meaningful effects on the relative availability and cost of EV and HEV products that can subsequently be used in the U.S.—particularly in States pursuing the California LEV II program.

3.3.1   Greenhouse Gas Registries and Reporting Programs

Over the last decade, various initiatives to register, document and promote voluntary GHG emission reduction measures have been introduced in the U.S. The goal of these programs is to encourage public and private entities to participate in GHG reduction activities and to test procedures for GHG emissions accounting. Each program afford individual project developers with the opportunity to register and document activities that help reduce GHG emissions and to possibly use the registered emission reductions for participation in a future emissions trading regime.

The different programs range in scope and project type, and do not all include activities related to transportation. However, three U.S.-based voluntary programs encourage developers of transportation projects to report the environmental benefits of their activities and submit project ideas: DOE's 1605(b) Voluntary Reporting of Greenhouse Gases Program, the California Climate Action Registry, and the U.S. Initiative on Joint Implementation (USIJI). Each of these programs is described below. Appendix 3 also lists several new and proposed State initiatives to register GHG emission reductions, many of which are designed to encourage the development of GHG reduction measures such as the increased use of EVs and HEVs.

U.S. Department of Energy's 1605(b) Voluntary Reporting of Greenhouse Gases Program

Managed by DOE's EIA, the 1605(b) Voluntary Reporting of Greenhouse Gases Program (created under Section 1605(b) of EPAct) affords any company, organization, or individual with the opportunity to establish a public record of their GHG emissions, emission reductions, and/or sequestration achievements in a central and public database. The program first began accepting reports on GHG reduction activities during calendar year 1995 and is thus among the world's first registries set up to track voluntary GHG activities.

Like other registries, 1605(b) lays the foundation for maintaining information about individual projects, and standardizing GHG emissions accounting methodologies, which in turn makes possible the creation of a market wherein GHG emission reduction credits can be traded. Participants generally participate in the program to gain recognition for environmental stewardship, demonstrate support for voluntary approaches to achieving environmental policy goals, support information exchange, and inform the general public about GHG reduction activities. If the participant has the emissions reductions certified by an independent third party entity, and the reductions meet the standards of a given emissions trading regime, then the participant may trade the certified credits within that regime and reap the financial benefits associated with the sale of those credits at market

---

[94] Id.

price. One example of such a regime, although still in its infancy, is the Chicago Climate Exchange, described in Section 3.3.2.

Data from the most recent 1605(b) reporting cycle, covering activities through 2000, were released by EIA in February 2002 and include considerable information on real-world transportation projects. Of the 72 transportation projects reported to the program, fifteen were EV projects involving emissions reductions of roughly 3,923 metric tons of carbon dioxide equivalent (CO_2E). Appendix 4 presents summary information on these projects, including the entities that undertook and reported the project, the name, scope and general description of each project, and the methods used to estimate the achieved GHG emission reductions. The data reported to the program is publicly available on DOE's website and may be useful for educational and project replication purposes.[95] For more information, contact the 1605(b) Program Communications Center at: 1-800-803-5182 or visit http://www.eia.doe.gov/oiaf/1605/frntvrgg.html.

The 1605(b) program will likely receive increased attention in the future, as the Bush Administration is redesigning and grooming it to be the main national system for tracking emissions and emission reduction activities, and perhaps establishing credits. Thus, the standards and methodologies that it establishes may become the default national standard that other registries and reporting programs, such as those mentioned below, will derive from and comply with it.

California Climate Action Registry

As established under State Senate Bill 1771, the State of California chartered in September, 2002, the California Climate Action Registry—a non-profit organization providing a central and standardized system for reporting annual GHG emissions reductions, including those reductions from motor vehicle activities. In return for voluntary registration of GHG emissions, the Registry promises to use its best efforts to ensure that participating organizations receive appropriate consideration under any future international, federal, or state regulatory regimes relating to GHG emissions.[96] Given the steps, described in Section 3.2.1, that California is taking to address vehicular GHG emissions, the registry may gain increased prominence for transportation related activities. For example, the bill directs the California Climate Action Registry to develop procedures for reporting and registering vehicular GHG reductions to the Registry.

In contrast to the 1605(b) program, entities participating in the California Registry have to report on *all* their emissions and emission reductions. At this point in time, the Registry does not accept reports that only include project-specific activities. Companies that wish to report on their transportation-related activities therefore also have to complete an inventory of company-wide emissions before submitting a report to the Registry.

For more information about the California Climate Action Registry, contact Diane Wittenberg, tel.213-891-1444; email: diane@climateregistry.org; or go to http://www.climateregistry.org/.

U.S. Initiative on Joint Implementation (USIJI)

The U.S. Initiative on Joint Implementation (USIJI) is designated as the official U.S. Government institution accepting jointly implemented GHG emission reduction projects

---

[95] See http://www.eia.doe.gov/oiaf/1605/frntvrgg.html for more information.
[96] California Energy Commission, Global Climate Change & California, http://www.energy.ca.gov/global_climate_change/index.html.

as part of the UNFCCC AIJ Pilot Phase.[97] A key goal of the USIJI program is to influence the technological choices associated with the already substantial private capital flows to developing countries. Any U.S. private sector firm, non-governmental organization, government agency, or individual is eligible to submit a project proposal to USIJI,[98] including developers of EV and HEV projects. The application criteria for participating in USIJI are outlined in Appendix 5. Proposals must be submitted in partnership with foreign host country participants, including any citizen or entity recognized by a host country, which has signed, ratified or acceded to the UNFCCC. To date, the USIJI has approved 50 projects in 26 countries,[99] and while no transportation projects have been approved, representatives of the program have indicated a particular interest in receiving projects that involve emission reductions from this sector.

At the time of writing this report, the status of USIJI has been put on hold pending a U.S. government review of how it fits within the Administration's overall climate change strategy.

3.3.2   Emerging Markets for Trading in GHG Credits

Another development, that is likely to have a significant impact on the development of GHG-related transportation projects and the increased market penetration of EVs and HEVs, is the emergence of a new market for trading in GHG emission reduction credits. Though few governments have imposed binding restrictions on GHG emissions, many companies have already begun exploring the benefits and challenges of GHG trading.[100] Even without government-imposed restrictions, emission reduction credits still have market value as long as there is a demand for the purchase of these credits. This demand exists and is steadily increasing, driven in part by the anticipation of one or more regulatory regimes, and by the desire to earn a reputation as an environmentally conscious entity. As a result, a small but growing market for the sale and transfer of credits based on GHG reduction activities has evolved over the past few years. As this market continues to grow, opportunities for selling and trading credits derived from GHG reduction activities in the transportation sector will also increase. Potential GHG reduction opportunities that could be generated and sold for credit on the GHG market include projects promoting the use of cleaner vehicle options, such as EVs and HEVs.

Trading activities have evolved in concert with a series of project-based mechanisms set up to gain experience and explore ways to address the climate change issue cost-effectively.   These programs and initiatives include USIJI; Ontario, Canada's multi-

---

[97] The UNFCCC introduced the concept of joint implementation (JI), which refers to arrangements through which an entity in one country partially meets its domestic commitment to reduce GHG levels by financing and supporting the development of a project in another country. To test the concept of JI, the Activities Implemented Jointly (AIJ) Pilot Phase was established at the first Conference of the Parties to the UNFCCC (COP-1), held in Berlin in 1995. Projects initiated during this phase were called "activities implemented jointly" to distinguish them from the fully-fledged JI projects the Convention may allow in the future. The goal of the AIJ Pilot Phase was to provide developing nations with advanced technologies and financial investment while allowing industrialized nations to fulfill part of their reduction commitment at the lowest cost. Because of the temporary pilot status of this program, it was decided that project developers cannot receive credit or other monetary incentives for projects developed and approved as part of this initiative.

[98] For further information on the USIJI program and project criteria, contact USIJI, 1000 Independence Avenue SW, Washington DC 20585, USA. Tel: (1-202) 586-3288, Fax: (1-202) 586-3485/3486.

[99] Project descriptions can be found at the USIJI website at www.gcrio.org/usiji/projects/CurrentProjs.html and at the AIJ website at www.unfccc.int/program/aij/.

[100] Only the United Kingdom and Denmark have established formal emissions trading programs as a component of domestic climate change policies. The European Union is developing the rules for an EU-wide GHG trading program, which is expected to enter into operation in 2005.

stakeholder Pilot Emissions Reduction program (PERT); [not a trading or credit program] the Dutch government's Emission Reduction Unit Procurement Tender (ERUPT); and the World Bank's Prototype Carbon Fund (PCF).

Since there is no central recording entity for tracking GHG emissions trades, the actual size of the market is not fully known. However, it is estimated that approximately 65 inter-company transactions have occurred since 1996, involving roughly 50 to 70 million metric tons of CO_2E emissions reductions.[101] This number may be conservative as several companies are reluctant to make their trading activities public. The price of these trades has ranged between \$0.60 and \$3.50 per metric ton of CO_2E. Most of these trades have been between buyers and sellers in Europe and North America. The majority of these trades have been verified by third party, independent entities.

The most popular trading activities have included fugitive gas capture from landfills, fuel switching, energy efficiency, and co-generation.[102] None of the trades have involved reductions from transportation activities, highlighting the lack of experience with generating project-based GHG emission reductions in the transportation sector. However, as it is fairly straightforward to monitor and demonstrate ownership for this type of reductions it is likely that the types of activities traded will expand to include emission reductions from transport projects.

Chicago Climate Exchange

The Chicago Climate Exchange is emerging as one of the key organizations for helping to generate a viable trading market for GHG emissions reduction credits. In June 2001, 33 companies with assets in the midwestern United States (including the Ford Motor Company) announced the formation of the Chicago Climate Exchange (CCX). Led by Environmental Financial Products and the Kellogg Graduate School of Management at Northwestern University, under a grant from the Joyce Foundation, the group will explore the potential for a regional GHG trading exchange in order to achieve a specified level of emission reductions. The companies have indicated in letters to CCX that they will consider trading on the exchange if effective rules are designed. The CCX has proposed that participating companies voluntarily commit to emissions reductions and trading in six GHGs.[103] Participants would commit to reducing their GHG emissions by 2 percent below 1999 levels by 2002 and reduce them 1 percent annually thereafter. Credits would be given for domestic and international emissions offsets projects after particular monitoring, verification, tracking and reporting requirements have been fulfilled. Potential emission reduction activities that could receive credit under the Chicago Climate Exchange include projects that reduce emissions from the transportation sector. Sample project types suggested by the CCX include fuel switching and vehicle efficiency improvement projects.

The CCX hopes to have the exchange up and running by the third quarter of 2002 for participants in seven states: Illinois, Indiana, Iowa, Michigan, Minnesota, Ohio, and Wisconsin. In 2003, the CCX aims to have commitments and trading among participants in the entire United States, Mexico and Canada, and to expand the exchange to include international participants in 2004.

---

[101] Richard Rosenzweig, Matthew Varilek, Ben Feldman, Radha Kuppalli, and Josef Jansen. The Emerging International Greenhouse Gas Market. PEW Center on Global Climate Change. Washington, DC. March 2002.

[102] Review and Analysis of the Emerging International Greenhouse Gas Market. Executive Summary of a confidential report prepared for the World Bank Prototype Carbon Fund. Natsource, 2001.

[103] The six gases covered by the CCX are carbon dioxide (CO_2), methane (CH_4), nitrous oxide (N_2O), hydrofluorocarbons (HFCs), perfluorocarbons (PFCs), and sulphur hexafluoride (SF_6).

For more information on the Chicago Climate Exchange contact: info@chicagoclimateX.com. Chicago Climate Exchange, 111 W. Jackson, 14th Floor, Chicago, Illinois 60604 USA. Phone: 1 (312) 554-3350, Fax: 1 (312) 554-3373, website: http://www.chicagoclimatex.com.

4 GHG Emissions From Battery-Powered Electric and Hybrid Electric Vehicles

4.1 Introduction

In developing and reporting on the GHG emission reductions associated with implementing EV and HEV activities, project developers should have a thorough understanding of the procedures for GHG emissions accounting. The following subsections provide an overview of the issues related to estimating and reporting on the potential GHG emission reductions achieved by replacing conventional internal combustion engine vehicles with EVs or HEVs. This section first discusses the types of domestic and international transportation-related GHG reduction projects that have been undertaken and for which data has been reported. Next is a discussion of the types and sources of GHG emissions associated with both EVs and HEVs, followed by an overview of some of the studies and models that may be helpful for estimating emission reductions from vehicle projects. Finally, this section briefly summarizes the most common rules and procedures for estimating and reporting on GHG emission reduction activities under project-based GHG mitigation programs.

4.2 Projects Deploying EV and HEV Technologies to Reduce GHG Emissions

There are five main types of activities that can be undertaken to reduce GHG emissions in the transportation sector. These include:

- *Changing vehicle fuel type:* sample activities include switching from gasoline/diesel to biodiesel, natural gas, electric batteries, fuel cells, and other alternative fuels;
- *Changing vehicle fuel efficiency:* for example, improving fuel economy, traffic management/infrastructure changes and/or vehicle scrappage programs;
- *Mode switching to less GHG-intensive transportation options:* increased public transportation, light rail systems, etc.;
- *Reducing transportation activity:* this could involve improved transit systems, road pricing, or telecommuting; and
- *Increasing vehicle occupancy rate:* activities may include car sharing, telematic systems for freight, or subsidized public transport.

Each option focuses on different ways to reduce emissions, ranging from behavioral changes to direct substitution of transport technologies. Hence, the procedures for estimating and accounting for emission reductions are different for each of the five activity types.

For the individual electric and hybrid electric vehicle project developer, the first option is the most relevant, as it refers to activities that can be undertaken directly by the individual fleet manager. For example, by replacing a fleet of one hundred gasoline-powered vehicles with one hundred electric battery-powered vehicles, a fleet manager can reduce GHG emissions by using a less GHG intensive fuel. The other four transportation activity types would mostly involve behavioral or regulatory changes that would likely be implemented by public authorities, automobile manufacturers, or private companies seeking to reduce the transport activities of their employees. Because this guide is targeted towards GHG reduction projects involving the deployment of electric and hybrid electric vehicles by individual fleet managers, we will focus on estimating emissions from vehicle technology and/or fuel switching projects.

There is little international experience with developing and implementing vehicle fuel switching projects specifically with the purpose of reducing GHG emissions.[104] Of the 157 projects registered with the UNFCCC Secretariat as AIJ pilot projects, only one takes place in the transportation sector.[105] This project, known as the RABA/IKARUS Compressed Natural Gas Engine Bus project, is funded by Dutch investors and hosted in Hungary. The project involves the development and testing of a new compressed CNG engine to be installed by the companies of RABA and Ikarus in new buses.[106] No EV or HEV projects have been reported to the UNFCCC Secretariat.

In the U.S., the number of voluntary actions to reduce GHG emissions in the transportation sector is also low. In 2000, there were 72 transportation related GHG emissions reduction projects reported to the DOE Voluntary Reporting of Greenhouse Gases Program—a small number compared to the 462 electricity generation, transmission, and distribution projects reported for the same year.[107] Nearly half (31) of these transportation projects involved AFVs and 15 involved the use of EVs. The AFV project developers reported an average estimated emissions reduction of 505 metric tons of CO_2E per project for 2000. (A more detailed description of these projects is provided in Appendix 4.) Concerned with the lack of transportation sector projects, national joint implementation offices have been promoting their development. For example, DOE issued a grant in the fall of 2000 to the Washington D.C.-based Center for Sustainable Development in the Americas (CSDA) to create an AIJ project using natural gas vehicles in Santiago, Chile.

4.3   GHG Emissions Associated with EVs and HEVs

The GHGs most closely identified with the transportation sector include CO_2, N_2O, and CH_4. Each GHG contributes differently to global warming, and this difference can be expressed by the global warming potential (GWP) of each gas. The GWP of a GHG is the degree to which that gas will enhance the overall effect of global warming. It is a function of the gas' direct or indirect radiative forcing potential (or how well the gas transmits visible radiation and traps infrared radiation). GWP is expressed in relative terms, with CO_2 as the base, for a given period of time. The concept of GWP allows us to compare the emissions of different GHGs, such as CH_4 and N_2O, using a common unit: kg of CO_2-equivalent (CO_2-E). GWPs recommended by the Intergovernmental Panel on Climate Change (IPCC) are

Table 4-1	Global Warming Potentials of Selected GHGs	
Greenhouse Gas	Global Warming Potential (100 Years)
Carbon Dioxide (CO_2)	1
Methane (CH_4)	21
Nitrous Oxide (N_2O)	310

---

[104] Most project-based GHG reduction activities target sectors such as electricity generation, industrial energy use, renewable energy development, or land use and forestry activities.

[105] http://www.unfccc.int/program/aij/aijproj.html.

[106] AIJ Uniform Reporting Format: Activities Implemented Jointly under the Pilot Phase. The RABA/IKARUS Compressed Natural Gas Engine Project, http://www.unfccc.int/program/aij/aijact/hunnld01.html.

[107] Energy Information Administration. http://www.eia.doe.gov/oiaf/1605/frntvrgg.html.

included in Table 4-1. In the case of EVs and HEVs, CO_2 is the major GHG emitted.

Both hybrid and battery-powered electric vehicles can result in considerable GHG emission reductions compared to conventional petroleum-fueled vehicles. However, the associated emission reductions vary, depending on the power generation mix used to charge the electric batteries and the fuel type used for fueling the hybrid vehicle. Because vehicle efficiency of EVs and HEVs is determined by several factors, including fuel type and propulsion system, no single value for potential emission reductions can be provided. Emission estimates will have to be determined for each vehicle model and fuel type used.

Figure 4-1   Changes in Fuel-Cycle GHG Emissions Relative to Gasoline Vehicles Fueled with Clean Gasoline[108]

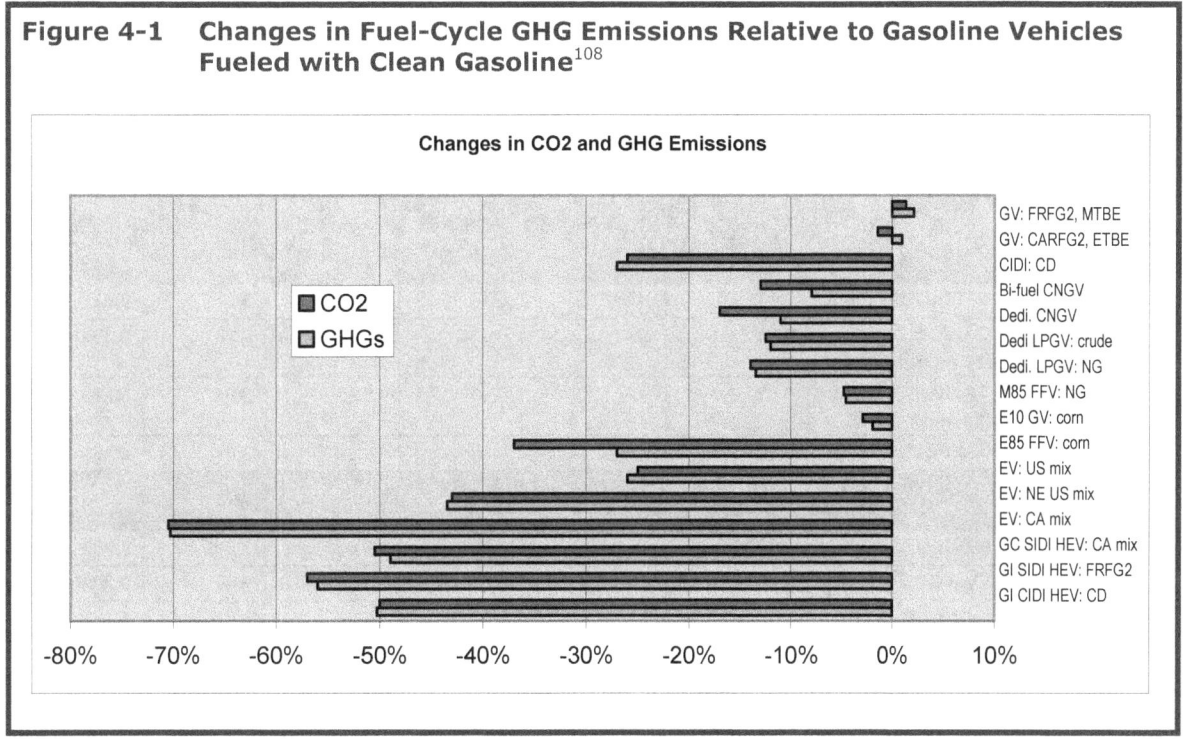

Figure 4-1 illustrates the potential changes in CO_2 and CO_2-equivalent emissions that can be obtained by replacing conventional gasoline vehicles with EVs and HEVs. GHG emissions are the sum of emissions of CO_2, CH_4, and N_2O, weighted by their GWPs, but, as Figure 4-1 illustrates, with the exception of ethanol and natural gas vehicles, CO_2 accounts for nearly all of the vehicles GHG emissions. The HEV options examined include grid-independent (GI) HEVs fueled with California gasoline (RFG2), grid-independent HEVs fueled by clean diesel, and grid-connected HEVs powered by a California electricity mix.   Grid-independent HEVs use the engine to recharge their batteries while grid-connected vehicles are required to be plugged into a stationary power

---

[108] Argonne National Laboratory. GREET 1.5—Transportation Fuel-Cycle Model. Volume 1: Methodology, Development, Use and Results. August 1999. GI=grid independent; CIDI=compression ignition, direct injection; FRFG2=Federal Phase 2 reformulated gasoline; SIDI=spark ignition, direct injection; E85=mixture of 85 % ethanol and 15% gasoline by volume; FFV=fuel flexible vehicle; E10=mixture of 10 % ethanol and 90% gasoline by volume; GV=gasoline vehicle; M85=mixture of 85 % methanol and 15% gasoline by volume; NG=natural gas; LPGV=liquefied petroleum gas vehicle; dedi=dedicated; CNGV=compressed natural gas vehicle; CD=conventional diesel; CARFG2=California Phase 2 reformulated gasoline; ETBE=ethyl tertiary butyl ether; MTBE=methyl tertiary butyl ether.

outlet to recharge the batteries. The EV options include an electricity mix, typical of three geographic regions around the country: California, the northeast U.S., and the total U.S. Other options include vehicles using: ethanol mixed with gasoline (E85—85% ethanol, and E10—10% ethanol), methanol mixed with gasoline (M85), liquid petroleum gas (LPG), and compressed natural gas (CNG).

To complement Figure 4-1, which shows GHG reductions as a percentage decrease compared to conventional gasoline vehicles, Table 4-2 lists typical CO_2 emissions associated with EVs and HEVs, as well as three conventional vehicles. It should be emphasized that the information presented in Table 4-2 are directly proportional to several factors that may have a wide degree of variation (such as vehicle efficiency and electricity fuel mix), and are therefore presented only to provide a relative order of magnitude.

Table 4-2	Typical CO_2 Emissions from Select Vehicle Options					
Vehicle	Vehicle Efficiency	Electrical Efficiency	Energy per Unit Fuel	Fuel CO_2 Emission Factor	C to CO_2	Vehicle Emission Factor
	gallon/mi	%	GJ/ gallon	kg C per GJ fuel	kg C per kg CO_2	kg CO_2/mi
CA: SUV (17 mpg)	0.06	NA	0.13	18.90	3.67	0.542
CA: AVE (26 mpg)	0.04	NA	0.13	18.90	3.67	0.354
CA: High-E (38 mpg)	0.03	NA	0.13	18.90	3.67	0.242
CA: HEV (55 Mpg)	0.02	NA	0.13	18.90	3.67	0.167
	KWh/mi	%	GJ/kWh	kg C per GJ fuel	kg C per kg CO_2	kg CO_2/mi
EV-Coal	0.30	35	0.00	26.80	3.67	0.303
EV-NGSS	0.30	30	0.00	15.30	3.67	0.202
EV-NGCC	0.30	45	0.00	15.30	3.67	0.135
EV-Hydro/RE	0.30	NA	NA	NA	3.67	0.000

Abbreviations:

CA = Commercial Automobile	High-E = High Efficiency	NGSS = Natural Gas Single Cycle
	HEV = Hybrid Electric Vehicle	NGCC = Natural Gas Combined Cycle
SUV = Sport Utility Vehicle	EV = Electric Vehicle	RE = Renewable Energy

The data listed in Table 4-2 assume a uniform electricity mix in the calculation of CO_2 emissions from EVs (i.e. the EVs listed are powered *either* by coal, natural gas, or hydro), when in fact there will be a variation of fuels, efficiencies and electricity mix for each region of the country. In Figure 4-1 above, the emission estimates are calculated based on the electric generation mix of California (CA), Northeast U.S. (NE US) and the national U.S. mix. The largest reductions occur for EVs with the California electric generation mix, where 48 percent of electricity is produced from hydropower plants. A more detailed discussion of the emissions from EVs and HEVs is provided in the following sections. In general, EVs and HEVs reduce GHG emissions by more than 40 percent, mainly because of their efficiency gains.

4.3.1 GHG Emissions From EVs

Because electric vehicles use batteries as the sole source of power generation, the procedures for measuring and estimating GHG benefits of EVs is different from hybrid electric and conventional vehicles. Battery-powered electric vehicles have no tailpipe emissions of GHGs and local air pollutants, but there are emissions associated with generating electricity for battery recharging. There are also some emissions associated with producing and scrapping the batteries. However, these emissions represent a small share relative to the total. Figure 4-2 shows that more than 90 percent of the GHGs emitted from EVs come from the process of producing, transporting, and storing fuel. The remaining GHG emissions are emitted during the feedstock-related stage, which includes feedstock recovery, transportation, and storage. No GHG emissions are associated with the vehicle operation stage, covering vehicle refueling and operations. Figure 4-3 is included for comparison, showing the share of fuel-cycle energy use and emissions of conventional gasoline vehicles by stage. Figure 4-3 shows that more than 80 percent of GHGs emitted from gasoline vehicles come from vehicle operation. Another 15 percent is emitted during the fuel production, transport, and storage stages.

Figure 4-2   Shares of Fuel-Cycle Energy Use and Emissions by Stage: Battery-Powered EVs[109]

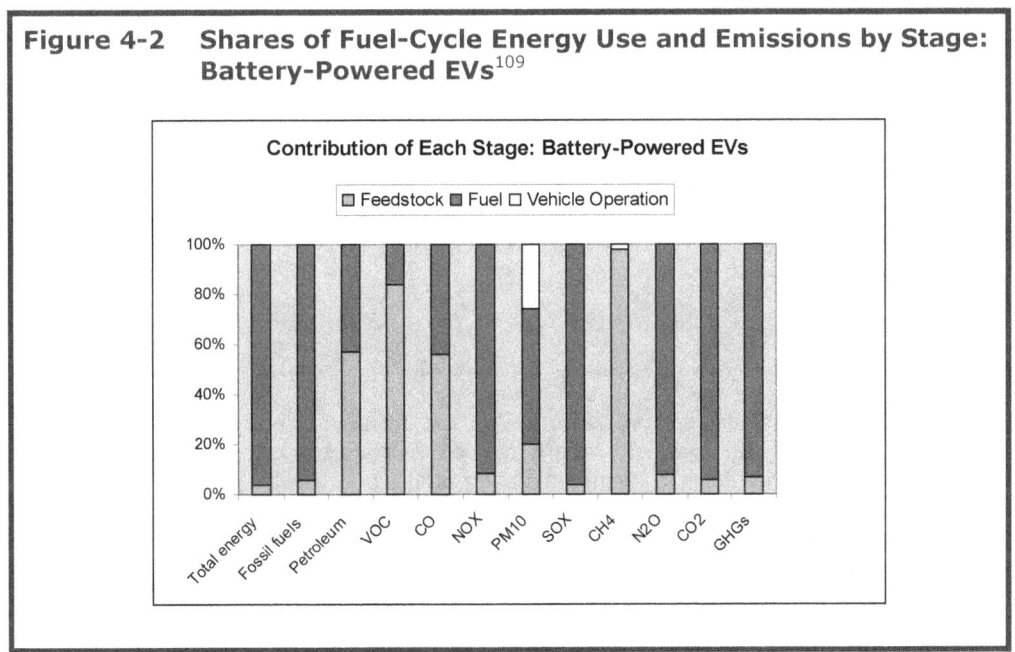

As most of the GHG emissions associated with the use of EVs are emitted during the downstream process, a full fuel cycle analysis of emissions should be used to estimate the GHG benefits of EV projects.

Depending on the power generation mix in use for the area where an electric vehicle is recharged, the overall emissions can be much less than those from conventional gasoline vehicles. If the battery of the electric vehicle is recharged in a region with a very coal intensive electricity generation mix, GHG emissions will be higher than if the battery is recharged with electricity from mainly renewable or natural gas-based electricity. EVs may have nearly zero total emissions when recharged with electricity generated by nuclear power or renewable sources. As illustrated in Figure 4-1 above, EVs recharged in California result in much lower emissions than EVs recharged in Northeastern U.S. The GHG benefits of California vehicles are even higher if compared with EVs recharged with electricity similar to the U.S. average fuel mix.

---

[109] Argonne National Laboratory. GREET 1.5—Transportation Fuel-Cycle Model. Volume 1: Methodology, Development, Use and Results. August 1999. The model assumes that the batteries are recharged with an average U.S. electric generation mix, under which 54 percent of electricity is generated from coal.

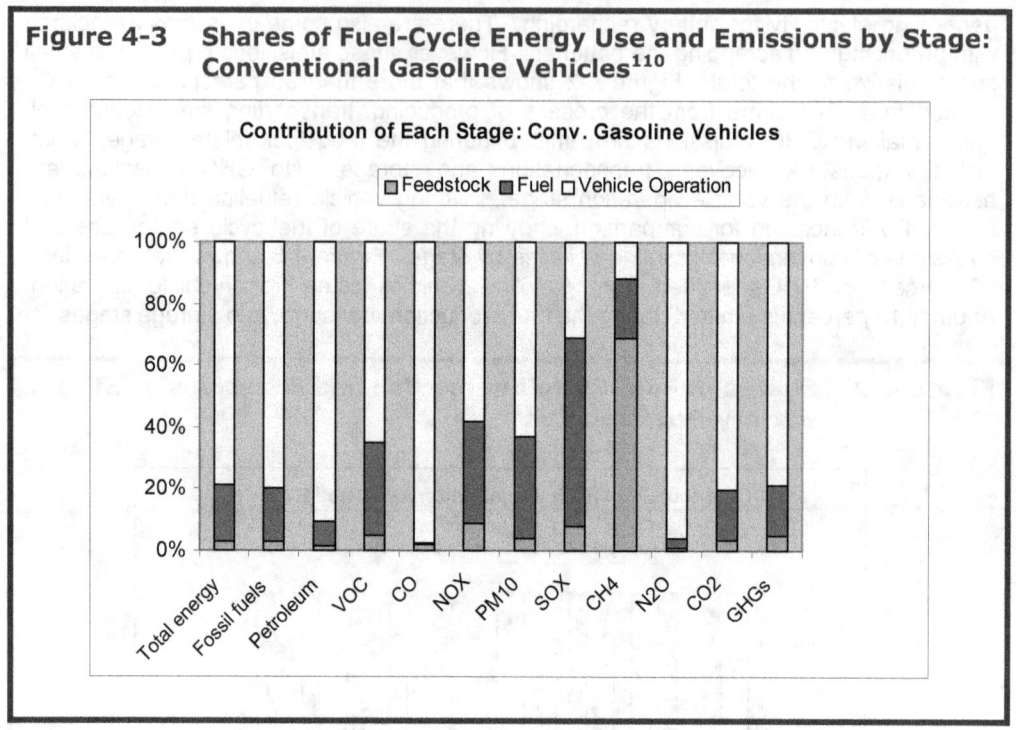

Figure 4-3   Shares of Fuel-Cycle Energy Use and Emissions by Stage: Conventional Gasoline Vehicles[110]

To establish the particular emissions from EVs, project developers must determine the electricity mix in the particular area where the batteries are being recharged. To assist project developers in the US with estimating emission benefits of GHG reduction activities, EIA has developed state average electricity emission factors that can be used to determine region-specific CO_2 emissions per unit electricity used.[111] (Specific emissions factors for each state and major city are listed in Appendix 6.) The EIA electricity emission factors were developed to provide reporters to the EIA 1605(b) Voluntary Reporting of Greenhouse Gases program with a means to convert their energy savings estimates into emission reduction estimates. Essentially, CO_2 emissions for all utility power plants over a three-year period were summed across all plants and then divided by the total generation of the plants to yield state-wide average emission factors.[112]

Project developers who wish to use more localized electricity emission factors—either for the purpose of higher accuracy or because they are located outside the US—should contact the local utility to obtain information on the electricity mix in the area where the EVs will be recharged. This information can then be used to derive local electricity emission factors (specified, e.g., in pounds of greenhouse gas (gas) per kWh), according to the following basic equation:

$$EEF = \sum_{i=1}^{n} [(FC_i) (FFEF_i)] \ / \ G_t$$

where:

EEF	= electricity emission factor (pounds gas per kWh)
FC$_i$	= total fuel consumed of type i (kcal, Btus or J)

---

[110] Id.

[111] Instructions for Form EIA-1605 Voluntary Reporting of Greenhouse Gases, 2000 (for data through 1999). EIA Energy Information Administration, US Department of Energy.

[112] EIA's emission factors were computed by multiplying fuel consumption data by fossil fuel emission factors—provided by IPCC—and dividing by electric generation.

4 GHG Emissions

| FFEF$_i$ | = fossil fuel emission factor for fuel type i (lbs. gas per kcal, Btus or J) |
| G$_t$ | = total generation across all fuel types |

The specifics of what should be included in the fuel consumption and generation variables will vary depending on the generation and fuel mix in the local area. However, these considerations do not affect the basic functional form of the equation, or the appropriate values for the fossil fuel emission factors.[113] Generally recognized fossil fuel emission factors, such as have been developed by IPCC and other recognized bodies, can applied in the above equation. The fossil fuel emission coefficients used by the EIA 1605(b) Voluntary Reporting Program are listed in Appendix 7 of this report.

Once local- or state-wide emission factors have been derived—typically expressed in terms of lbs of CO_2E per kWh—the vehicle owner can then determine emissions of an individual vehicle type by multiplying the vehicle efficiency (kWh/mile) by the emission factor (lbs of CO_2E per kWh). After emissions per distance traveled have been established for the electric vehicle, a comparison can be made with the emissions per distance traveled of a conventional gasoline vehicle to determine the actual emission reductions attributable to the EVs. The emission factor for gasoline, listed in Appendix 7, is estimated at 19.564 lbs of CO_2 per gallon.

4.3.2 GHG Emissions From HEVs

HEVs run on conventional fuels, and thus the vast majority of their GHGs are emitted from the tailpipe during vehicle operation, as is the case with conventional vehicles. Figure 4-4, above, shows the share of GHG emissions during each stage of the HEV fuel use cycle, using grid-independent HEVs fueled by gasoline.[114] The figure illustrates that at least 80 percent of GHGs emitted are released during the vehicle operation stage. Similar to the case with gasoline vehicles, another 15 percent of GHGs are emitted during the fuel production, transport, and storage stages. Advanced HEVs, such as fuel cell vehicles using hydrogen, are like EVs in that some or all of their GHG emissions may be produced during the fuel production and distribution stages.

4.4 Studies and Measurements of GHG Emission Benefits of EVs and HEVs

As GHGs are only regulated in a few countries, only a limited number of studies and publicly available resources are available to offer assistance in estimating GHG emissions from vehicles. The following summaries provide an overview of the major information sources on emissions benefits from EVs and HEVs.

---

[113] In the above equation, the fuel consumption and fossil fuel emission factors should always be expressed in energy units (calories, Btus, or joules) rather than physical units (e.g., tons, barrels, or cubic feet). As long as this basic rule is followed, the uncertainty surrounding the fossil fuel emission factors will be minimized because within any given fuel type, the variability in the relationship between heat content and carbon content is very limited. In general, the heat content of a given fuel depends primarily on the fuel's carbon content. There are other chemical components of fuels that also contribute to the fuel's heat content including sulfur and, most notably, hydrogen. However, these other chemical constituents tend to represent only a small fraction of the fuel's chemical makeup relative to carbon. Hence the variations in the ratio of carbon to heat content, for a given type of fuel, are likewise small. And since a fuel emission factor is simply the ratio of the fuel's carbon to heat content, the uncertainty range surrounding the factor is small.

[114] Argonne National Laboratory. GREET 1.5—Transportation Fuel-Cycle Model. Volume 1: Methodology, Development, Use and Results. August 1999.

Figure 4-4   Shares of Fuel-Cycle Energy Use and Emissions by Stage: Grid Independent HEVs, Internal Combustion Engine (ICE) with Reformulated Gasoline[115]

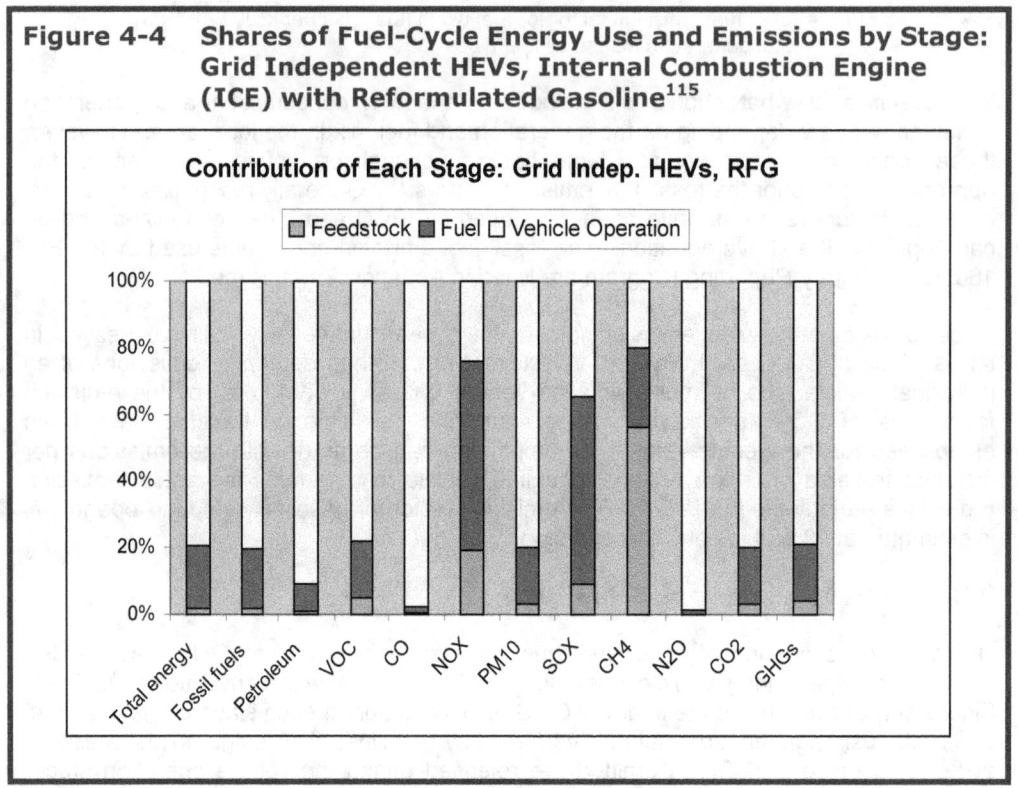

4.4.1   Greenhouse Gases, Regulated Emissions, and Energy Use in Transportation (GREET)

GHG emissions for vehicles are easily calculated using the Greenhouse Gases, Regulated Emissions, and Energy Use in Transportation (GREET) Model. The GREET model was developed by the Argonne National Laboratory to make calculations of the GHG emissions of light duty conventional vehicles and alternative fueled vehicles in the U.S. All the GHG emissions from vehicle use and upstream from fuel production, are included. Three GHGs (CO_2, N_2O and CH_4) are combined with their GWPs to calculate CO_2-E GHG emissions. GREET also evaluates criteria pollutant emissions, and compares fuel efficiency and emissions for EVs and HEVs relative to conventional gasoline vehicles. Users are able to change the default values to accommodate their specific situation. The GREET model is free of charge and can be downloaded from http://www.transportation.anl.gov/ttrdc/greet.

It should be emphasized that the model is based on U.S. conditions and energy infrastructure. Users from other countries should be careful to adopt model inputs, which are relevant to country-specific conditions. These should include country-specific assumptions regarding fuel use and GHG emissions during the production, refining, and transportation of fuels and the national electricity mix used for electricity generation.

---

[115] Argonne National Laboratory. August 1999. Total GHG emissions also include fuel recovery, production, and distribution to the vehicle. However, the emissions associated with these steps are relatively small compared to the emissions emitted during operation of the vehicle.

4.4.2   Canada's Transportation Climate Change Table

In May 1998, Canada's federal, provincial, and territorial Ministers of Transportation established the Transportation Climate Change Table as part of the national process to develop a climate change strategy.[116] The Table was comprised of transportation sector experts from a broad cross-section of business and industry, government, environmental groups and non-governmental organizations. It was mandated to identify specific measures to mitigate GHG emissions from Canada's transport sector.

The Transportation Climate Change Table submitted its Options Paper, "Transportation and Climate Change: Options for Action" to the Ministers of Transportation and the National Climate Change Secretariat in November 1999. The Options Paper assesses the costs, benefits and impacts of over 100 measures. The Transportation Table undertook 24 studies in support of the Options Paper. As part of one of these studies, several alternatives to gasoline, including EVs, were compared for their potential for reductions in GHG emissions.[117] Table 4-3 summarizes the findings regarding the relative energy efficiency, GHG emissions, and costs of alternative fuels.

As can be seen from the "Net GHG Ratio" column of Table 4-3, EVs emit considerably fewer GHGs than conventional vehicles. However, review of the information provided in the study indicates that the GHG benefits projected for EVs are based on the assumption that relatively little of the electricity is generated from the combustion of fossil fuels, as is typically the case in Canada. This assumption may not be appropriate for most areas in the U.S. As illustrated in Appendix 6, some U.S. states are heavily reliant on coal for their electricity generation. Hence, in some areas of the U.S., the GHG benefits could be less dramatic depending on the fuel mix of the particular state in question.

The second column of the table is labeled "Relative Vehicle Efficiency". The values in the column represent the distance traveled per BTU of fuel consumed for each fuel relative to gasoline. The third column of the table, labeled "Upstream GHG Ratio," shows estimates of how the GHG emissions associated with production of alternative fuels and alternatively fueled vehicles compare to gasoline and gasoline-fueled vehicles. In the case of EVs, GHG emissions from the upstream process is higher (1.08) than the emissions from conventional gasoline vehicles (1.00). This is because most of the emissions from EVs are produced during the production of electricity for the batteries, whereas emissions from gasoline vehicles are emitted during operation of the vehicle itself and are therefore not included in this column. The fourth column of Table 4-2 is labeled "Net GHG Ratio", reflecting the combined effects of "upstream" GHG emissions differences and changes in GHG emissions per unit distance of vehicle travel. Once both vehicle operation (tail pipe emissions) and upstream processes are combined, EVs are shown to emit far fewer GHGs (0.30) than conventional gasoline vehicles (1.00) (with the assumed electricity producing fuel mix). The fifth column of Table 4-3 is labeled "Gasoline Equivalent Fuel Cost" and represents the relative price of the alternative fuel per unit of energy. The sixth and seventh columns in Table 4-3 contain the "Vehicle Price Change" for 2010 and 2020. These columns show the change in Retail Price Equivalent

---

[116] For more information on Canada's Transportation Climate Change Table visit the website at: http://www.tc.gc.ca/envaffairs/english/climatechange/ttable/ or email: TCCTable@tc.gc.ca.

[117] Alternative and Future Fuels and Energy Sources For Road Vehicles. Prepared for Canada's Transportation Issue Table, National Climate Change Process. Levelton Engineering Ltd. in association with (S & T)² Consulting Inc., BC Research Inc., Constable Associates Consulting Inc., Sierra Research.
http://www.tc.gc.ca/envaffairs/subgroups/vehicle_technology/study2/Final_report/Final_Report.htm.

(RPE)[118] compared to conventional vehicles associated with vehicles designed to use the alternative fuel.

Fuel Type	Relative Vehicle Efficiency	Upstream GHG Ratio	Net GHG Ratio	Gasoline Equivalent Fuel Cost	Vehicle Price Change 2010	Vehicle Price Change 2020
Table 4-3	**Efficiency, Emissions, and Cost Comparisons for Alternative Fuels Relative to Conventional Gasoline**					
Conventional Gasoline	1.00	1.00	1.00	1.00	NA	NA
RFG	1.00	1.08	1.01	1.04	$0	$0
E10 (corn)	1.01	0.96	0.97	1.12	$0	$0
E10 (cellulose)	1.01	0.88	0.95	1.12	$0	$0
E85 (cellulose)	1.11	-1.00	0.36	1.85	$0	$0
M85	1.11	1.01	0.89	1.56	$0	$0
LPG	1.10	0.52	0.74	1.09	$1,000	$750
CNG	1.10	0.66	0.75	0.78	$2,300	$1,000
Fuel Cell (M100 NG)	1.61	0.81	0.62	1.22	$7,000	$4,000
Fuel Cell (H2 NG)	2.05	1.72	0.48	2.51	$7,000	$4,000
Fuel Cell (H2 Elec.)	2.05	1.84	0.51	2.51	$7,000	$7,000
EV	[119] 3.71	1.08	0.30	0.54	$10,000	$8,000
Diesel	1.41	0.72	0.78	0.71	$2,310	$2,310
Diesel 50ppm S	1.41	0.76	0.79	0.78	$2,310	$2,310

4.4.3 The Environmental Protection Agency (EPA) MOBILE6 Model

The emission rates of local air pollutants of AFVs and engines are readily available from EPA. The EPA also has a model (MOBILE6), which allows fleets to calculate the emissions reductions they can expect in real-world operation when using AFVs. MOBILE6 is a computer program that estimates HC, CO, and NO_x emission factors for gasoline and diesel fueled highway motor vehicles, as well as for AFVs such as natural gas and electric vehicles that may be used to replace them. MOBILE6 calculates emission factors for 28 individual vehicle types in low- and high-altitude regions of the United States. MOBILE6 emission factor estimates depend on various conditions, such as ambient temperatures, travel speeds, operating modes, fuel volatility, and mileage accrual rates. Many of the variables affecting vehicle emissions can be specified by the

---

[118] The RPE is defined as the average "retail price equivalent" that must be achieved for the manufacturer to remain economically viable over the longer term. In this report RPE accounts for fixed costs (e.g., engineering, facilities, tooling), variable costs (e.g., purchased parts, assembly labor), manufacturer markup (to cover overhead and profit), and dealer margin. The formula can be described as follows: RPE = ((Fixed Cost/Unit + Variable Cost/Unit) * Mfr. Markup) * Dealer Markup.

[119] The energy efficiency assigned to electric vehicles in this analysis does not account for the inefficiency associated with generating electricity from the combustion of fossil fuels. Previous analysis has shown that the efficiency advantage for electric vehicles will be less pronounced if the electricity used for vehicle recharging is generated from fossil fuel-fired power plants.

4 GHG Emissions

user, tailoring the calculations to specific types of fleets. MOBILE6 will estimate emission factors for any calendar year between 1952 and 2050, inclusive. Vehicles from the 25 most recent model years are assumed to be in operation in each calendar year by MOBILE6. Some states, such as California, have similar software which are specific to their unique climate and driving characteristics. Estimates of emissions reductions are often needed for AFV owners to apply for and receive grants from incentive programs.

EPA is undertaking an effort to develop the next generation of modeling tools for the estimation of emissions produced by on- and off-road mobile sources, which includes the New Generation Model. The design of this modeling system is guided by four broad objectives: (a) the model should encompass all pollutants (including HC, CO, NO_x, PM, air toxics, and GHGs) and all mobile sources at the levels of resolution needed for the diverse applications of the system; (b) the model should be developed according to principles of sound science; (c) the software design should be efficient and flexible; and (d) the model should be implemented in a coordinated, clear, and consistent manner. EPA views the New Generation Model as a logical next step in the continuing effort to improve mobile source emissions models to keep pace with new analysis needs, new modeling approaches, and new data.

4.5 Procedures for Estimating GHG Emissions Benefits from EV and HEV Projects

During the past decade, a series of project-based programs have been introduced to gain experience and harness the power of markets in order to address the issue of climate change in a cost-effective manner. Each of these programs is governed by a unique set of rules. However, they exhibit some common elements that constitute a *de facto* (though non-binding) set of minimum quality criteria that govern the creation of credible emission reductions. Leading examples of these programs and initiatives include USIJI; the AIJ Pilot Phase, Canada's Pilot Emissions Reduction Trading Program (PERT) in Ontario; Oregon's Climate Trust; the Emission Reduction Unit Procurement Tender (ERUPT) of the Dutch government; and the World Bank Prototype Carbon Fund (PCF).

The following rules and procedures are common elements of project-based systems. They provide a framework for project developers interested in developing GHG reduction projects.

4.5.1 GHG Emissions Baseline

The emissions baseline is an integral part of the GHG reduction project proposal as the baseline is used to estimate emissions benefits of the project and will be used as the basis for awarding credits to the project. Many project-based programs measure emission reductions by comparing the emissions performance of a credible "without project" baseline against the "with project" emissions.

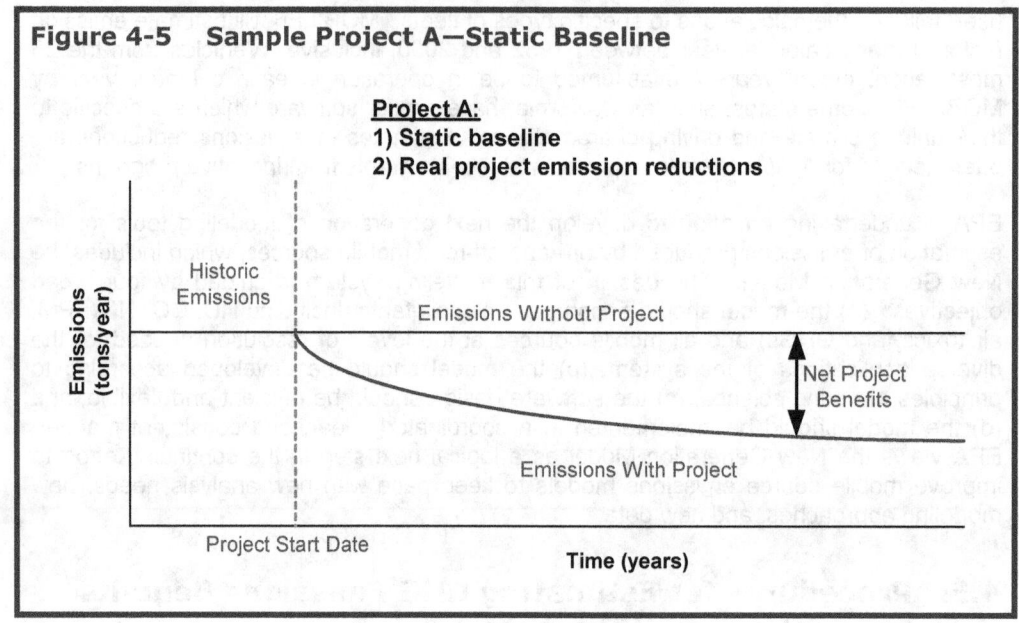

Figure 4-5   Sample Project A—Static Baseline

The baseline typically refers to the process of developing an understanding of the "without project" emissions performance scenario—either static or dynamic—which can then be used for comparison with the emissions of the project. The challenge of developing emission baselines stems from the problem of projecting what will happen in a given economy 10, 20, or 30 years down the line. Static baselines rely on historical information to fix emissions at a set level, such as an entity or project's physical emissions in a given year. This same emissions level is then maintained every year throughout the life of the project. An example of a static emission baseline is provided in Figure 4-5.

Dynamic baselines are emission baselines that try to take into account changes that are likely or expected to happen during the life of the project. As such, dynamic baselines are linked to particular variables and may be revised upward and downward depending on project and entity characteristics such as output levels, growth rates, efficiency rates, and peer group benchmarks. For example, changes that could happen include future laws mandating use of a similar technology or fuel option, increased demand for transportation leading to increased vehicle usage, expected commercialization of a similar vehicle type, and so on. If such changes are taken into account, it will no longer be sufficient to use historical data for deriving the "without project" scenario, and some assumptions and adjustments regarding emissions levels in future years will have to be made. Figure 4-6 illustrates what a dynamic baseline might look like. It should be noted that dynamic baselines do not always involve an increase in emissions. In some cases, a general adoption of cleaner and more efficient technologies may lead to lower emissions in the "without project" scenario. In this case, the dynamic baseline would have a downward sloping curve and credits would only be awarded to projects that improve the emissions performance even further.

Regardless of which baseline scenario is selected, project developers must be careful to describe all assumptions used and explain exactly why a particular methodology is utilized.

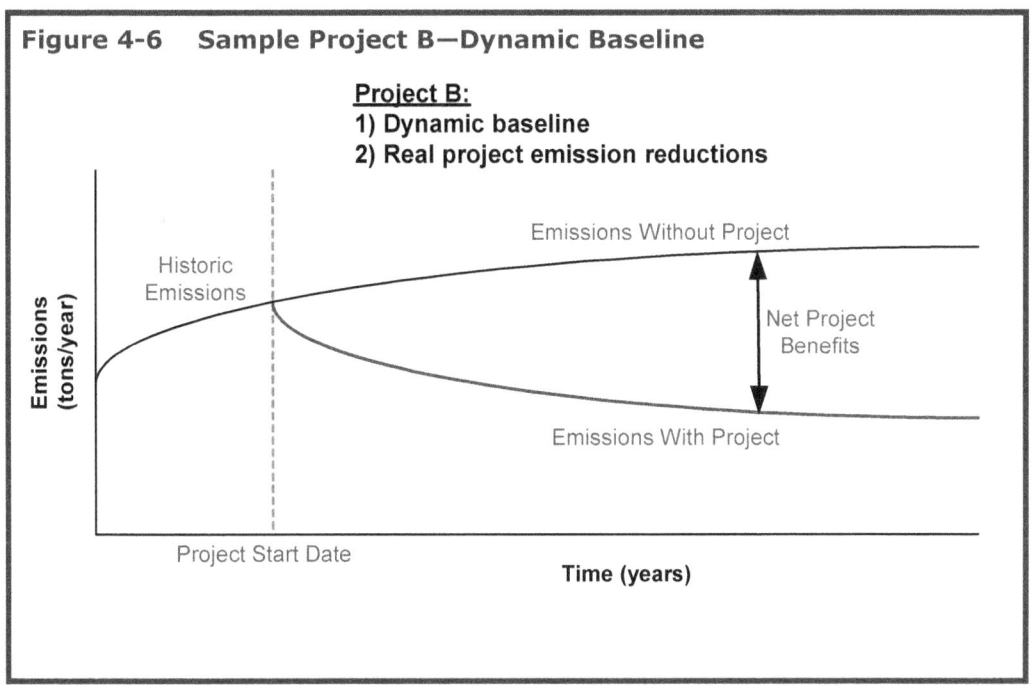

Figure 4-6   Sample Project B—Dynamic Baseline

Project B:
1) Dynamic baseline
2) Real project emission reductions

Emissions (tons/year)

Historic Emissions

Emissions Without Project

Net Project Benefits

Emissions With Project

Project Start Date

Time (years)

Once the baseline has been determined, the estimate of emissions "with the project" can be developed. To determine project emissions the same assumptions and time frames used for the "without project" baseline should be applied. Most project cases lead to *real* emission reductions. However, as illustrated in Figure 4-7, it is sometimes possible that emissions "with the project" will continue to rise above historical emissions. Such projects may still be able to obtain GHG reduction credits, as long as the reported project emissions performance continues to fall below the emissions associated with the baseline scenario.

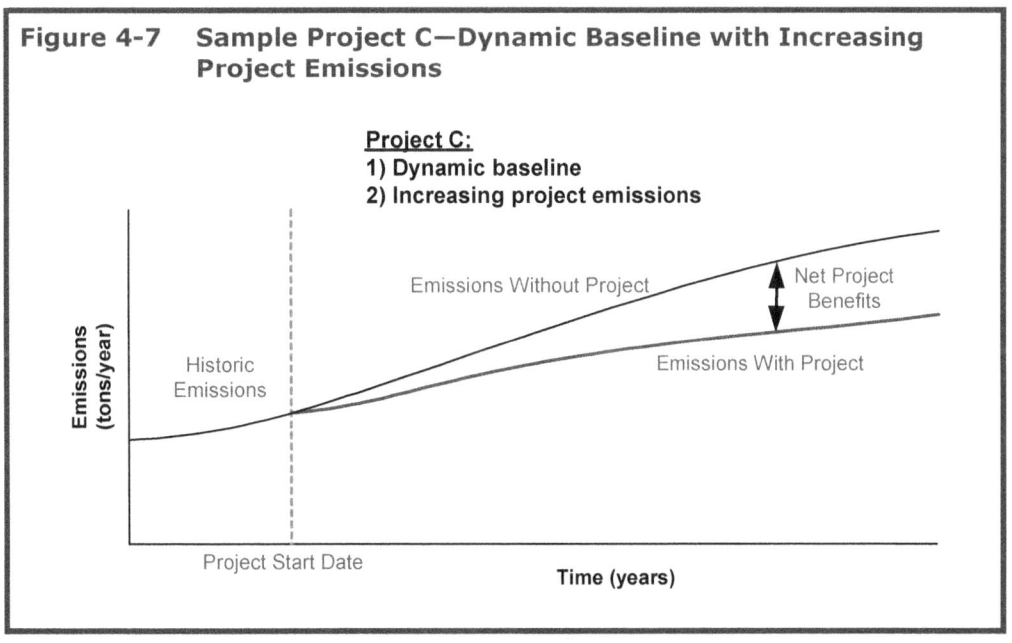

Figure 4-7   Sample Project C—Dynamic Baseline with Increasing Project Emissions

Project C:
1) Dynamic baseline
2) Increasing project emissions

Emissions (tons/year)

Historic Emissions

Emissions Without Project

Net Project Benefits

Emissions With Project

Project Start Date

Time (years)

In the case of EV and HEV projects, the emission baseline would be calculated by looking at the distance traveled and/or fuel use of both the new vehicles and the vehicles to be replaced.

To date, no GHG crediting program has developed specific guidance for estimating GHG benefits from vehicle projects. However, some useful indicators for how to calculate emissions can be derived from the WRI/WBCSD Greenhouse Gas Protocol Initiative, which is a multi-stakeholder initiative between industry, government, and non-governmental organizations, to develop generally accepted accounting practices for measuring and reporting corporate GHG emissions.[120] The resulting standard and guidance are supplemented by a number of user-friendly GHG calculation tools, which can be accessed on the GHG Protocol website (www.ghgprotocol.org).[121] Although the GHG Protocol focuses on corporate emissions, the proposed accounting standards and reporting instructions serve as an indicator of how project-specific emission reductions could be calculated.[122]

According to the GHG protocol, there are two general methodologies for calculating emissions from vehicle projects: fuel-based and distance-based.[123] The preferred method, known as the "fuel-based" approach, is based on previously aggregated fuel consumption data to determine emissions. Following this approach, fuel consumption is multiplied by the CO_2 emission factor for each fuel type in order to derive CO_2 emissions. The fuel emission factor is developed based on the fuel's heat content, the fraction of carbon in the fuel that is oxidized, and the carbon content coefficient. To calculate emissions the following equation should be used:

$$CO_2 \text{ Emissions} = \text{Fuel Used} \times \text{Heating Value} \times \text{Emission Factor}$$

In the case that project developers do not have access to site-specific information, default emission factors and heating values for different transportation fuels are listed in the guidance for using the GHG Reporting Protocol.

Fuel use data can be obtained from several different sources including fuel receipts, financial records on fuel expenditures, or direct measurements of fuel use. If specific information on fuel consumption is not available, information on vehicle activity data (i.e. distance traveled) and fuel economy factors (such as miles per gallon) can be used to calculate fuel consumption, using the following equation:

---

[120] *The Greenhouse Gas Protocol: A Corporate Accounting and Reporting Standard.* WRI/WBCSD. Washington, D.C. 2000. Under the GHG Protocol, corporate transportation emissions take the form of either direct or indirect emissions. Direct emissions refer to emissions that are associated with owned or controlled sources, such as company owned vehicle fleets and corporate aircraft. Indirect emissions refer to all other company-related emissions, including employee commuting, short-term vehicle rentals, and upstream/downstream transportation emissions. If companies purchase electricity for owned or operated EVs, the related emissions should be reported as indirect emissions and should use guidance developed in the 'Stationary Combustion Tool' for calculating emissions. For all other vehicles, including HEVs, companies should use the methodologies developed for calculating direct emissions from mobile sources.

[121] Only transportation-related CO_2 emission estimates are included in this tool. According to the GHG Protocol, accounting for N_2O and CH_4 emissions is optional at the discretion of the user. This is because N_2O and CH_4 emissions comprise a relatively small proportion of overall transportation emissions.

[122] The WRI/WBCSD is also in the process of developing a GHG project accounting model with the aim of developing general guidance for emission reduction and land use, land-use change and forestry projects. This module will include accounting procedures for transportation projects.

[123] *"Calculating CO_2 Emissions from Mobile Sources"* WRI/WBCSD GHG Protocol Initiative. www.GHGprotocol.org.

4 GHG Emissions

> Fuel Use = Distance x Fuel Economy Factor

The GHG Protocol also includes default fuel economy factors for different types of mobile sources and activity data.

The second methodology, the "distance-based" approach, should only be used in case information on fuel use cannot be obtained. In the distance-based method, emissions are calculated by using distance based emission factors to calculate emissions. Activity data could be expressed in terms of vehicle-kilometers (or miles) traveled, passenger-kilometers (or miles), and so on. This information is then multiplied by a default distance-based emission factor[124] according to the following equation:

> CO_2 Emissions = Distance Traveled x Distance-Based Emission Factor

Default distance-based emission factors are provided in the guidance for the GHG Protocol.

The distance-based approach is less accurate than the fuel-based approach, and is thus recommended as the last resort for corporate GHG accounting purposes. As accuracy is an extremely important issue in terms of developing and crediting GHG reduction projects, the fuel-based approach is also the preferred approach for project-specific GHG reduction activities.

4.5.2   Environmental Additionality

The requirement of environmental additionality is linked closely to the process of developing the GHG emissions baseline. The concept of environmental additionality refers to the notion that the emission reductions achieved by the project must be proven not to have occurred in the absence of the project. That is, it is important that the credits awarded to the project developers must stem from emission reduction activities undertaken *in addition to* the business-as-usual scenario. Otherwise, the credits claimed from the project will not result in true, long-term environmental benefits, and the project developers will be awarded credits for emission reductions that never really took place. Hence, a major part of the additionality criterion involves proving that emission reductions were not a result of general technology improvements or activities undertaken to comply with existing regulations. To be credible, baselines should therefore take into account any laws, regulations, or technology improvements that may have a direct or indirect impact on GHG emissions.

In the case of EVs, the question of additionality is pretty straightforward due to the general lack of EV market penetration and limited prospects for increased market penetration in the near future. Because of the limited use of EVs, it is pretty safe to argue that the purchase of such vehicles would not have happened without the specific EV project in question—unless of course the purchase of such vehicles were mandated by an existing law or regulation. However, in the case of HEVs it becomes a little more tricky to argue for the additionality of a potential HEV project. In some countries, such as Japan and the U.S., HEVs have reached a very limited market penetration. In these countries, the existing level of market penetration should be accounted for in the "without

---

[124] A sample default distance-based emission factor could be 0.28 kg CO_2 per mile traveled for a small petrol car with no more than a 1.4 liter engine. *"Calculating CO_2 Emissions from Mobile Sources"* WRI/WBCSD GHG Protocol Initiative. www.GHGprotocol.org.

project" emissions scenario, unless the project developer is able to clearly demonstrate that such vehicles would not have been purchased by the individual fleet manager in the absence of the HEV project.

4.5.3 Leakage

Another common criterion requires that the project developers provide evidence that the emissions reductions achieved at the project site do not lead to increases in emissions outside the boundaries of the project (i.e., emissions "leakage"), or that the calculation of claimed emissions reductions quantifies and accounts for leakage. Switching to electric vehicles is a good example of a project type with potential for leakage. If the boundary of the project is limited to an analysis of tailpipe emissions alone, the emissions will be reduced to zero, when in fact significant emissions may be produced at the power plant in the generation of the electricity for powering the electric vehicle. These power plant emissions would have "leaked" from the accounting system.

4.5.4 Monitoring and Verification

Another common requirement is that project developers develop a plan or procedures for how emission reductions are monitored throughout the life of the project. The measured reductions must then be verified by an independent third party, who certifies that monitored reductions and/or the proposed method for calculating emissions performance can be or has been audited to provide a credible quantitative assessment of actual project performance. Both the monitoring and verification requirements involve guidelines for validating and verifying that no leakage will take place and that the GHG emissions baseline is estimated correctly (i.e. that the reductions meet the environmental additionality requirement).

4.5.5 Ownership

Finally, most programs require that the project proponent has a legitimate claim to ownership of the reductions generated by the project and that other potential claimants are identified. Ownership can be demonstrated through documents certifying and dividing ownership clearly among all project participants. If necessary, supporting documents by local or national government authorities can be included to verify the validity of claimed ownership.

The issue of ownership is an important consideration for transportation projects. In many countries, buses and taxis are owned by individual vehicle operators rather than one single fleet operator. When the ownership of a transportation project covering 200 vehicles is divided among a similar number of owners, contractual and other issues may become very complicated. One solution may be to form an association representing all the vehicle owners, which could then be listed as the owner of the project.

4 GHG Emissions

5 Case Study on Quantifying GHG Emissions from Battery-Powered Electric Vehicles

5.1 Introduction

The following case study is based on a hypothetical project that involves the deployment of 125 electric battery charged taxis to replace 125 gasoline-fueled taxis. The case study focuses on the process of developing an emissions baseline and estimating net GHG emission benefits of an individual project.

The following subsections provide a brief summary of the project case study, outline the general criteria for developing a GHG reduction project based on current market-based proposals for GHG control, develop the project based on these criteria, and estimate the emissions baseline and net project benefits. Three sample baseline scenarios are provided to illustrate how different project characteristics may influence the baseline estimate. The three baselines include: (1) a static baseline assuming that the 125 new electric vehicles are purchased instead of 125 *new* conventional gasoline powered vehicles; (2) a dynamic baseline assuming that the 125 new electric vehicles will replace 125 *aging* conventional gasoline vehicles with an estimated average life time of eight years; and (3) a static *full fuel cycle* baseline, including fuel production and refining along with vehicle operation, that assumes that the 125 new electric vehicles are purchased instead of 125 *new* conventional gasoline powered vehicles.

5.2 Emission Reduction Project for Taxis

This case study is based on a hypothetical project in a country called the Clean Cities Republic.[125] Although the Clean Cities Republic is a developing country, it does not represent any country or region in particular. It should be emphasized that the numbers used for this case study are fictional. The data provided for estimating the emissions baseline have been developed to illustrate how to quantify potential emission benefits. The data should not be used as an indicator of the specific emissions potential of an electric vehicle project. Electric vehicle project developers should obtain their own GHG emission data for both the conventional vehicles to be replaced and the new alternative fuel vehicles to be introduced.

5.2.1 Republic of Clean Cities Background Information

The Republic of Clean Cities is a country with a population of 45 million people. Gross domestic product (GDP) is US$190 billion per year, with an annual growth rate of 5 to 6 percent over the last 10 years. As a result of this economic expansion, the country is

---

[125] The hypothetical country example of the Republic of Clean Cities was first introduced at the 6th National Clean Cities Conference for illustrating a similar case study on estimating the GHG benefits of a natural gas vehicle project. Julie Doherty and Jette Findsen, "Case Study: CNG Taxis, The Republic of Clean Cities," Presentation for the NETL-sponsored training session, *Developing International Greenhouse Gas Emission Reduction Projects Using Clean Cities Technologies*, in San Diego, CA, May 10, 2000.

experiencing an energy demand growth of 7 percent per year, with the transportation sector representing the fastest growing energy sector. Currently, transportation activities account for 32 percent of energy related CO_2 emissions, although this share is expected to grow significantly over the next few decades as the transportation sector continues to expand.

The project will be located in the capital of the Republic of Clean Cities, which is a city of 8 million people with a population growth of 5 percent per year. On average, there are 7 people per motor vehicle, compared to 1.3 per vehicle in the U.S. The total number of vehicles on the road is growing by 7 percent annually. The capital is experiencing serious local environmental pollution problems and is among the 20 most polluted cities in the world. The concentration of total suspended particulates (TSP) in the air is 8 times higher than the proposed World Health Organization (WHO) standards. The majority of the capital's pollution problems are caused by transportation emissions. To alleviate some of these environmental problems, the government has introduced tax incentives for switching to alternative fueled vehicles. In addition, a recently passed law mandates that all new cars should drive on unleaded gasoline. Currently, 40 percent of all gasoline sold in the country is leaded. The local government has also introduced a car use reduction plan to curb the rapid growth of new vehicles in the capital area. Finally, a new domestic regulation was put in place this year for reductions in vehicle tailpipe emissions of criteria pollutants.

To date, no electric vehicles have been purchased in the capital and there are no domestic manufacturers or dealers supplying electric vehicles.

5.3   The Project Case Study

As part of the project, 125 dedicated electric vehicles (sedans) will be purchased to either replace 125 existing or new conventional gasoline taxis of a similar size. To develop a supporting infrastructure, vehicle accessible electrical outlets will be provided at the site where these taxis are parked, including at the homes of the taxi drivers. Moreover, an extensive training course will be provided for the fleet mechanics. The lifetime of the project is estimated conservatively, at 12 years. Each taxi is expected to drive an average of 70,000 miles per year. The energy use of the electric vehicles is 1.46 kwh/mile and the mileage of the conventional gasoline vehicles that would have been purchased in the absence of the project is 26 miles per gallon of gasoline.

The project participants include the Capital City Transportation Department, a local taxi fleet operator, and a U.S.-based electric vehicle manufacturer. The electric vehicle project has been approved by the Republic of Clean Cities' National Climate Change Office, which has been authorized by the Ministries of Foreign Affairs, Energy, and Environment, to evaluate and certify internationally sponsored GHG reduction projects. The National Climate Change Office, administered by the Ministry of Environment, has provided written documentation of project approval.

The project reduces CO_2 emissions by reducing the need for oil recovery, gasoline refining, and fuel transportation, which produces more CO_2 emissions than recharging the electric batteries. The carbon intensity of electricity generated in the capital region is relatively low, as more than 35 percent of the generating capacity comes from hydropower. The remaining electricity is generated from a mix of coal and diesel. A comparison of N_2O and CH_4 emissions will not be included in the emissions baseline because these do not significantly contribute to projected emissions.

5.4 Project Additionality

Determining the additionality of the EV project is relatively straightforward. As mentioned earlier, there are no electric vehicles in the capital and the technology is not yet commercially available on the domestic market.   One major impediment for the introduction of EVs is the considerable higher cost of the vehicles and the lack of knowledge about the technology. Although tax incentives are provided for owners of AFVs there are no laws or regulations requiring public or private vehicle fleet owners to purchase alternative fuel vehicles, such as EVs. It is therefore unlikely that electric vehicles will be introduced in the country in the near future. The EV project is clearly additional and would likely qualify for credit under any market based GHG reduction program.

If, on the other hand, a law were in place mandating that 15 percent of all public and private fleets must consist of zero emission vehicles, such as electric vehicles, the issue of additionality would be less straightforward. In this case, the project developer would not be able to claim GHG emission reduction credits for EVs purchased to meet the 15 percent requirement. Only vehicles purchased to exceed the mandated zero emission requirements would receive credit. Hence, a fleet owner with 200 conventional gasoline taxis—who replaces 40 old conventional gasoline vehicles with 40 new EVs—would only be able to obtain emission reduction credits for 10 of the new EVs. The other 30 vehicles would go towards meeting the 15 percent mandate for zero emission vehicles. However, for the purposes of the following case studies it is assumed that no such laws have been put in place.

5.5 Estimating the Emissions Baseline

Since the introduction of the concept of cooperatively implemented GHG reduction projects, little experience has been gained regarding the development and evaluation of transportation-related GHG reduction projects.   As mentioned earlier, only one transportation project has been approved under the UNFCCC's AIJ Pilot Phase. One project, however, does not provide enough precedent to be used for the development of standardized methodologies for analyzing transportation projects.

For this type of project it should be sufficient to include information about CO_2 emissions only, instead of covering all GHGs, simply because CO_2 emissions account for most of the GHG emissions from the project. Clearly, the analysis should include a comparison of upstream emissions because the project emissions are dependent on the fuel mix used for generating the electricity used in the batteries.

This case study will provide three sample baseline scenarios to illustrate how different project characteristics may influence the baseline estimate. The three baselines include:

1. A static baseline assuming that the 125 new electric vehicles are purchased instead of 125 new conventional gasoline powered vehicles. These vehicles are purchased to meet growing demand for taxi services.
2. A dynamic baseline assuming that the 125 new electric vehicles will replace 125 aging conventional gasoline vehicles with an estimated average lifetime of eight years.
3. A static baseline assuming that the 125 new electric vehicles are purchased instead of 125 new conventional gasoline powered vehicles. This analysis includes a full fuel cycle analysis similar to that provided in the GREET model.

The purpose of presenting these different baseline scenarios is two-fold. One is to advance the discussion on some of the issues that must be resolved in order to establish clear guidelines for the documentation and approval of transportation-related projects. The other purpose is to provide potential project developers with an idea of the issues that must be considered during the development of an emissions baseline for a transportation project. Project developers can then choose between or combine the different baseline scenarios depending on the purpose and requirements of the program to which the project participants will be applying for credit.

Factors that may determine the choice of baseline scenarios, include:

1. The transportation technology used for the project;
2. Availability of full fuel cycle and tailpipe emissions data;
3. Individual GHG program requirements;
4. The risk tolerance and level of accuracy desired by project developers and investors; and
5. The acceptable level of transaction costs.

The three baseline scenarios are outlined in the following subsections. Each version of the baseline scenarios involves three quantification steps. These include a calculation of: (1) the project reference case, (2) project-related emissions, and (3) net emission benefits of the project.

The first quantification step entails estimating what the emissions would have been without implementing the project. This step is also known as the emission baseline or the project reference case and should include data for the entire life of the project. Because the potential project emission benefits are derived by comparing project emissions to the reference case, accuracy in the development of the reference case is very important. However, estimating future emissions is a difficult process. It is almost impossible to factor in everything that may or may not happen 10 to 20 years down the road. Moreover, many different results can be achieved depending on which assumptions are used to derive the future emissions scenario. GHG reduction programs and project developers planning to receive credit for their projects under a future market-based GHG reduction program must be careful to develop baseline criteria that would be stringent enough to be accepted under any program. Given the differences between the various initiatives to credit GHG reduction activities, developers should consult the preliminary guidelines of each of the proposed programs before developing a project, and be careful to detail all assumptions and emission sources when quantifying the potential emission benefits. The examples provided in the following case study are less comprehensive and should only be used as an indicator of the types of data and quantification procedures that could be required from the different GHG reduction programs.

The third quantification step involves estimating emissions from the project itself. The data provided should include an estimation of all relevant project emissions throughout the life of the project. During this process, project developers should be careful to define the boundary of the project and detail all the assumptions and emission sources included in the estimate.

The fourth and final quantification step is rather simple. It entails calculating the net benefits of the project. To derive the net benefits, the project developer must subtract the project emissions from the emissions estimated for the reference case. The difference will represent the net benefits of the project.

5.5.1 Emission Baselines: Version 1

The first scenario is based on a static emissions baseline. This means that the current level of business as usual emissions are assumed to remain constant throughout the life of the project. This scenario does not take into consideration changes that may occur over the life of the project such as declining vehicle efficiency or improvements in new vehicle technology.

In version 1, the method used to calculate emission reductions is based on a comparison of fuel usage and the corresponding fuel emissions factors.

Step 1: The Reference Case

The reference case represents what would happen if the GHG reduction project were not implemented. In this case, it is assumed that without the GHG reduction project, 125 new conventionally fueled gasoline vehicles would have been purchased to satisfy the growing demand for taxi services. Because version 1 of the case study assumes that emissions of the project are static, the GHG emissions rate of each taxi is assumed to remain the same over the next 12 years.

In this version of the case study, the formula for calculating emissions of the gasoline vehicles is:

> Emissions over the project lifetime = (miles driven per year) / (vehicle efficiency in miles per gallon, mpg) x (emission factor of gasoline) x (number of vehicles) x (number of project years)

The emission factor for gasoline is assumed to be 19.564 lbs CO_2/gallon (8.873 kg CO_2/gallon).[126] Hence emissions without the project would have been:

Project lifetime emissions = (70,000 miles) / (26 mpg) x (8.873 kg CO_2/gallon) x (125 vehicles) x (12 years)

= **35,833 metric tons CO_2 over the 12 year project lifetime**

Step 2: The Project Case

The project case represents the actual emissions of the project itself. In this instance, the project case refers to the emissions of the 125 electric vehicle taxis over the 12-year life of the project.

In this version of the case study, the formula for calculating emissions of the electric vehicles is:

> Emissions over the project lifetime = (miles driven per year) x (vehicle efficiency in kWh per mile) x (emission factor of electricity generation in kg CO_2 per kWh) x (number of vehicles) x (number of project years)

The emission factor for electricity generation in the capital area is assumed to be 0.178 kg CO_2/kWh. Hence emissions with the project would be:

Project lifetime emissions (over 12 years) = (70,000 miles) x (1.46 kWh per mile) x (0.178 kg CO_2/kWh) x (125 vehicles) x (12 years)

= **27,287 metric tons CO_2**

---

[126] U.S. Department of Energy, Energy Information Administration (EIA), Instructions for the Voluntary Reporting of Greenhouse Gases Program.

Step 3: Deriving Net Project Benefits

The net project emission benefits are derived by subtracting the project case from the reference case. As illustrated below, the net project benefits of version 1 of the case study are 8,545 metric tons of CO_2.

Reference case - project case = Net project benefits

35,833 - 27,287 = **8,545 metric tons of CO_2**

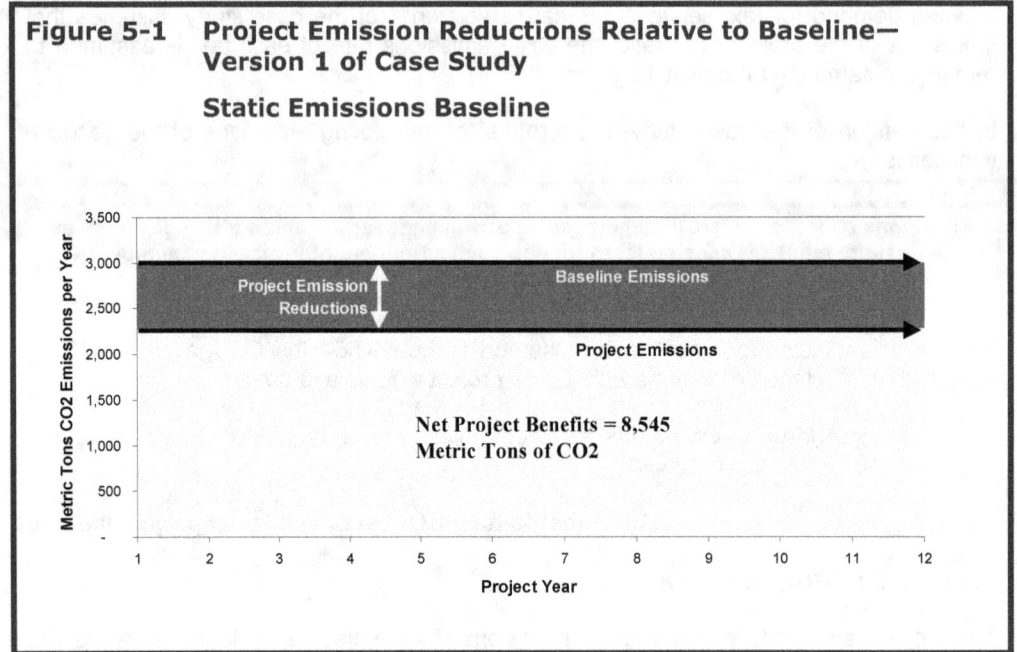

Figure 5-1   Project Emission Reductions Relative to Baseline— Version 1 of Case Study

Static Emissions Baseline

5.5.2   Emission Baselines: Version 2

The second scenario for the electric vehicle project relies on a dynamic emissions baseline. A dynamic baseline takes into account the changes that may happen to emissions and equipment as the vehicles age over time. In this version of the case study, it is assumed that the 125 new electric vehicles will replace an equal number of aging gasoline vehicles. However, as the old vehicles only have an estimated average lifetime of 8 years left, it must also be assumed that a similar number of new gasoline vehicles would be purchased after 8 years to replace the old vehicles as they are taken out of service. Therefore, in this version of the case study the mileage of the old gasoline vehicles is assumed to be considerably lower than the mileage of the new gasoline vehicles that are projected to be purchased 8 years into the future.

Step 1: The Reference Case

The reference case represents what would happen if the GHG reduction project were not implemented. As this is a dynamic baseline that takes into account the fact that the old gasoline vehicles are expected to be taken out of service after an average of 8 years— and be replaced with new gasoline vehicles—the reference case will be calculated in two steps. First, the emissions of the old vehicles during the first 8 years of the project are calculated, then the emissions of the new vehicles used during the last 4 years of the project lifetime will be estimated. The two numbers are then added together and will represent the emissions of the reference case. It is assumed that the mileage of the old gasoline vehicles is 21 gallons per mile while the mileage of the new vehicles will be 28 miles per gallon. It should be noted that additional layers of "dynamics" could be added, such as estimating the mileage of each vehicle from year to year to account for declining efficiencies as the vehicles age. However, for the sake of simplicity and a diminishing return on accuracy, a simple two tiered assumption is used: one tier for eight years of the old vehicles, and one tier for four years of the new vehicles.

The formula for calculating emission reductions is the same as in version 1 of the case study:

> Emissions over the project lifetime = (miles driven per year) / (vehicle efficiency in miles per gallon, mpg) x (emission factor of gasoline) x (number of vehicles) x (number of project years)

The emission factor for gasoline is assumed to be 19.564 lbs CO_2/gallon (8.873 kg CO_2/gallon).[127] Hence emissions without the project would have been:

Emissions (old vehicles; 8 year time scale) = (70,000 miles) / (21 mpg) x (8.873 kg CO_2/gallon) x (125 vehicles) x (8 years) = 29,589 metric tons CO_2

Emissions (new vehicles; 4 year time scale) = (70,000 miles) / (28 mpg) x (8.873 kg CO_2/gallon) x (125 vehicles) x (4 years) = 11,096 metric tons CO_2

Emissions of all gasoline vehicles over 12 year project lifetime = 29,589 + 11,096

= 40,685 metric tons CO_2

Step 2: The Project Case

The project case represents emissions of the project itself. In this situation, the project emissions remain the same as version 1 of the case study. Hence, project emissions are 27,287 metric tons CO_2.

Step 3: Deriving Net Project Benefits

The net project emission benefits are derived by subtracting the project case from the reference case. As illustrated below, the net project benefits of version 2 of the case study are 13,398 metric tons of CO_2 equivalent.

Reference case - project case = Net project benefits

40,685 - 27,287 = **13,398 metric tons of CO_2**

---

[127] U.S. Department of Energy, Energy Information Administration (EIA), Instructions for the Voluntary Reporting of Greenhouse Gases Program.

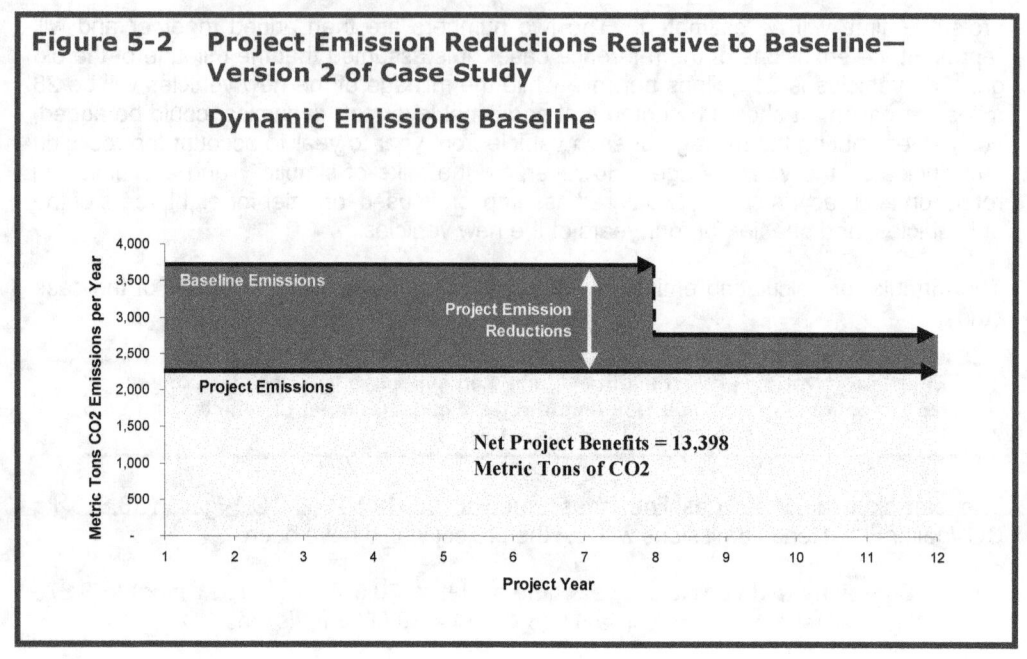

Figure 5-2   Project Emission Reductions Relative to Baseline—Version 2 of Case Study

Dynamic Emissions Baseline

5.5.3   Emission Baselines: Version 3

The third version of the emission baseline for the electric vehicle project relies on a static emission baseline, but uses a different model for quantifying the emissions benefits. In the previous two versions of the case study, emissions benefits were estimated by comparing fuel usage of the different vehicle types. However, this method does not account for the entire emissions scenario of the project. A more accurate analysis of emission benefits would analyze the entire project life cycle, including emissions from the production, transportation, processing, and combustion of the fuel used. However, this type of analysis is very complicated and would be costly to undertake for the individual project developer.

For projects in the United States, it would be possible to undertake this type of analysis by using the GREET model developed by Argonne National Laboratory.[128] As part of this model, emissions have been computed for a number of different vehicle types and models based on a detailed analysis of the energy production and usage of the entire transportation sector. Project developers can apply data regarding a specific vehicle model to the GREET model and calculate the potential GHG and other emissions reductions from a project. However, this model only applies to the transportation sector in the U.S. No similar studies have been undertaken in other countries. In particular, developing countries lack the adequate data and resources to undertake such studies of life cycle emissions.

The following version of the case study applies hypothetical electric vehicle data to the GREET model to illustrate how emissions would be calculated using this model. The baseline in this case study is more detailed than the two previous versions; that is, emissions data is presented for three stages of the fuel cycle. These stages include feedstock (production, transportation, and storage of primary energy feedstock), fuel

---

[128] Michael Wang, "Greenhouse Gases, Regulated Emissions, and Energy Use in Transportation (GREET)". Argonne National Laboratory. **www.transportation.anl.gov/ttrdc/greet**.

(production, transportation, storage and distribution of energy source), and vehicle operation (fuel combustion or other chemical conversion).

Step 1: The Reference Case

The reference case represents what would happen if the GHG reduction project were not implemented. As in the first version of this case study, it is assumed that 125 conventional gasoline taxis would have been purchased instead of the electric vehicles. The assumptions for the annual emissions from one gasoline vehicle, based on hypothetical data, are as described in Table 5-1.

Table 5-1	Version 3 of Case Study—Annual CO_2 Emissions Without the Project (grams/mile/year)		
Feedstock	**Fuel**	**Vehicle Operation**	**Total**
19	86	402	**507**

Emissions over 12 yrs = 0.507 kg CO_2 /mi x 70,000 mi x 125 cars x 12 yrs

= **53,249 metric tons of CO_2**

Step 2: The Project Case

As in the previous versions of this case study, the project case refers to the emissions of the 125 electric vehicle taxis over the 12-year life of the project. It is assumed that emissions of the electric vehicles will remain constant over the life of the project. Hence, a static baseline is used. The assumptions for the annual emissions from one electric vehicle, based on hypothetical data, are as described in Table 5-2.

Table 5-2	Version 3 of Case Study—Annual CO_2 Emissions with the EV Project (grams/mile/year)		
Feedstock	**Fuel**	**Vehicle Operation**	**Total**
21	237	0	**258**

Emissions over 12 yrs = 0.258 kg CO_2 /mi x 70,000 mi x 125 cars x 12 yrs

= **27,097 metric tons of CO_2**

Step 3: Deriving Net Project Benefits

The net project emission benefits are derived by subtracting the project case from the reference case. As illustrated below, the net project benefits of version 3 of the case study are 26,152 metric tons of CO_2.

Reference case - project case = Net project benefits
53,249 - 27,097 = **26,152 metric tons of CO_2**

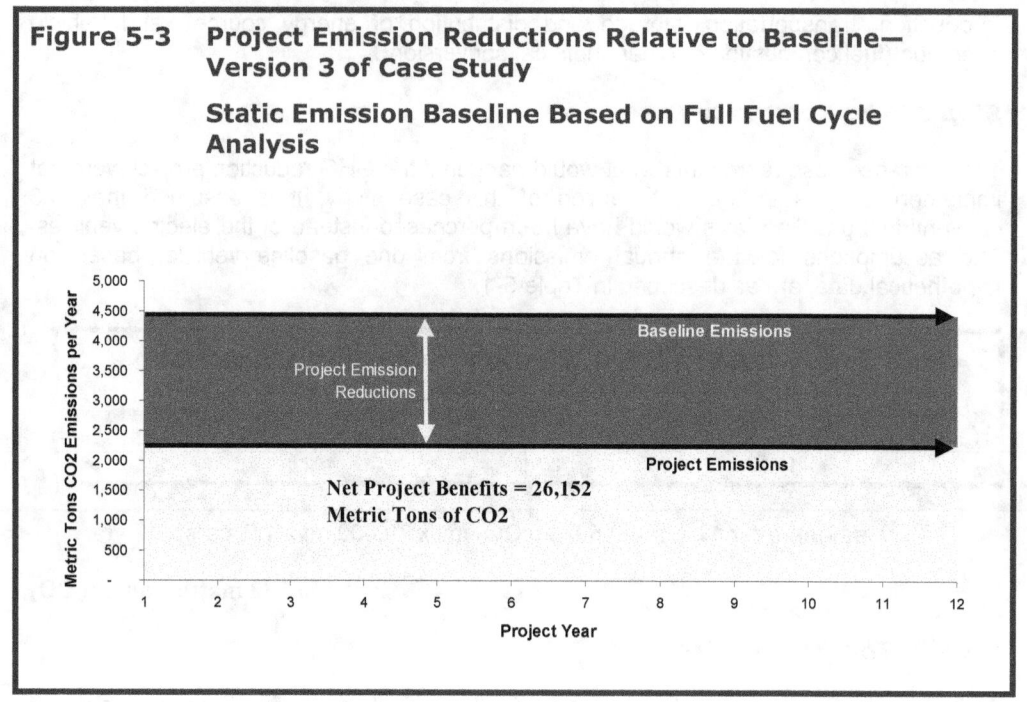

Figure 5-3   Project Emission Reductions Relative to Baseline—Version 3 of Case Study

Static Emission Baseline Based on Full Fuel Cycle Analysis

5.6 Discussion

The previous section presented three different methods for estimating the emission benefits of an EV project. Each of the three baselines represents a viable means of calculating emissions reductions resulting from the project. Ultimately, the specific circumstances of a potential EV or HEV project will determine which methodology is being used.

In general, there is a tradeoff between the accuracy of a baseline and the cost and effort associated with its calculation. As transportation projects are typically smaller in size and thus reap relatively few GHG emission reduction credits, it will be harder to justify the transaction costs involved with developing a very detailed estimate of projected emissions. As a result, project developers may prefer to use a less stringent baseline estimation procedure. However, as a general rule, project developers should aim to be as conservative as possible when determining emission reduction credits. Investors looking to purchase emission reduction credits want to ensure that the credits purchased are credible and minimize the risk of default in future commitment periods. Hence, they prefer to invest in credits that are based on sound and credible estimation procedures. It is therefore important that project developers clearly describe the baseline methodology and assumptions used, and explain why this approach was preferred over other methods. Moreover, in cases where there might be some uncertainty regarding the exact amount of expected emissions benefits—for example due to an expected decline in EV efficiency which cannot yet be quantified because of little experience with the technology—project developers should select the least optimistic emissions scenario. This type of estimation procedure is much more likely to gain acceptance by current and future GHG crediting programs.

6 Summary and Conclusions

Accurate and verifiable emission reductions are a function of the degree of transparency and stringency of the protocols employed in documenting project- or program-associated emissions reductions. The purpose of this guide is to provide a background for law and policy makers, urban planners, and project developers working with the many GHG emission reduction programs throughout the world to quantify and/or evaluate the GHG impacts of EVs and HEVs.

In order to evaluate the GHG benefits and/or penalties of EV or HEV projects, it is necessary to first gain a fundamental understanding of the technology employed and the operating characteristics of these vehicles, especially with regard to the manner in which they compare to similar conventional gasoline or diesel vehicles. Therefore, the first two sections of this paper explain the basic technology and functionality of battery-powered electric and hybrid electric passenger vehicles, but focus on evaluating the models that are currently on the market with their similar conventional counterparts, including characteristics such as cost, performance, efficiency, environmental attributes, and range.

Since the increased use of EVs and HEVs, along with alternative fuel vehicles in general, represents a public good with many social benefits at the local, national, and global levels, they often receive significant attention in the form of legislative and programmatic support. Some states mandate the use of EVs and HEVs, while others provide financial incentives to promote their procurement and use. Furthermore, Federal legislation in the form of mandatory standards or incentive programs can have a significant impact on the EV and HEV markets. In order to implement effective legislation or programs, it is vital to have an understanding of the different programs and activities that already exist, so that a new project focusing on GHG emission reduction can successfully interact with and build on the experience and lessons learned of those that preceded it.

Finally, most programs that deal with passenger vehicles—and with transportation in general—do not address the climate change component explicitly, and therefore there are few GHG reduction goals that are included in these programs. Furthermore, there are relatively few protocols that exist for accounting for the GHG emissions reductions that arise from transportation and, specifically, passenger vehicle projects and programs. These accounting procedures and principles gain increased importance when a project developer wishes to document in a credible manner, the GHG reductions that are achieved by a given project or program. Section four of this paper outlines the GHG emissions associated with EVs and HEVs, both upstream and downstream, and section five goes on to illustrate the methodology, via hypothetical case studies, for measuring these reductions using different types of baselines.

Unlike stationary energy combustion, transportation related emissions from some HEVs and EVs come from dispersed sources and require different methodologies for assessing GHG impacts. This resource guide outlines the necessary context and background for those parties wishing to evaluate projects and develop programs, policies, projects, and legislation aimed at the promotion of HEVs and EVs for GHG emission reduction.

Appendices

A1 Comparison of Electric and Hybrid-Electric Vehicles to Similar-Performance Gasoline-Powered Vehicles

This appendix compares an electric vehicle and a hybrid-electric vehicle to a similarly performing gasoline-powered vehicle. While the text discusses these different technologies in depth, this table format shows side-by-side comparisons of vehicles currently available for purchase or resale. This information was gathered from manufacturers and DOE vehicle emissions class ratings.

In this case, the comparison is between the battery-powered GM EV-1, the hybrid-electric Toyota Prius, and the gasoline-powered Honda Civic DX Sedan. All three vehicles have similar body specifications, but contrast in engine types. The major differences are found in performance measures of fuel efficiency, emissions, and range. The hybrid-electric Prius has an particularly notable fuel efficiency, because its city efficiency (52 mpg) exceeds its highway efficiency (45 mpg). In another contrast to conventional vehicles, the range of the hybrid-electric vehicle is also greater for city driving than for highway driving. The electric vehicle boasts no tailpipe emissions and holds the highest emissions rating at ZEV.

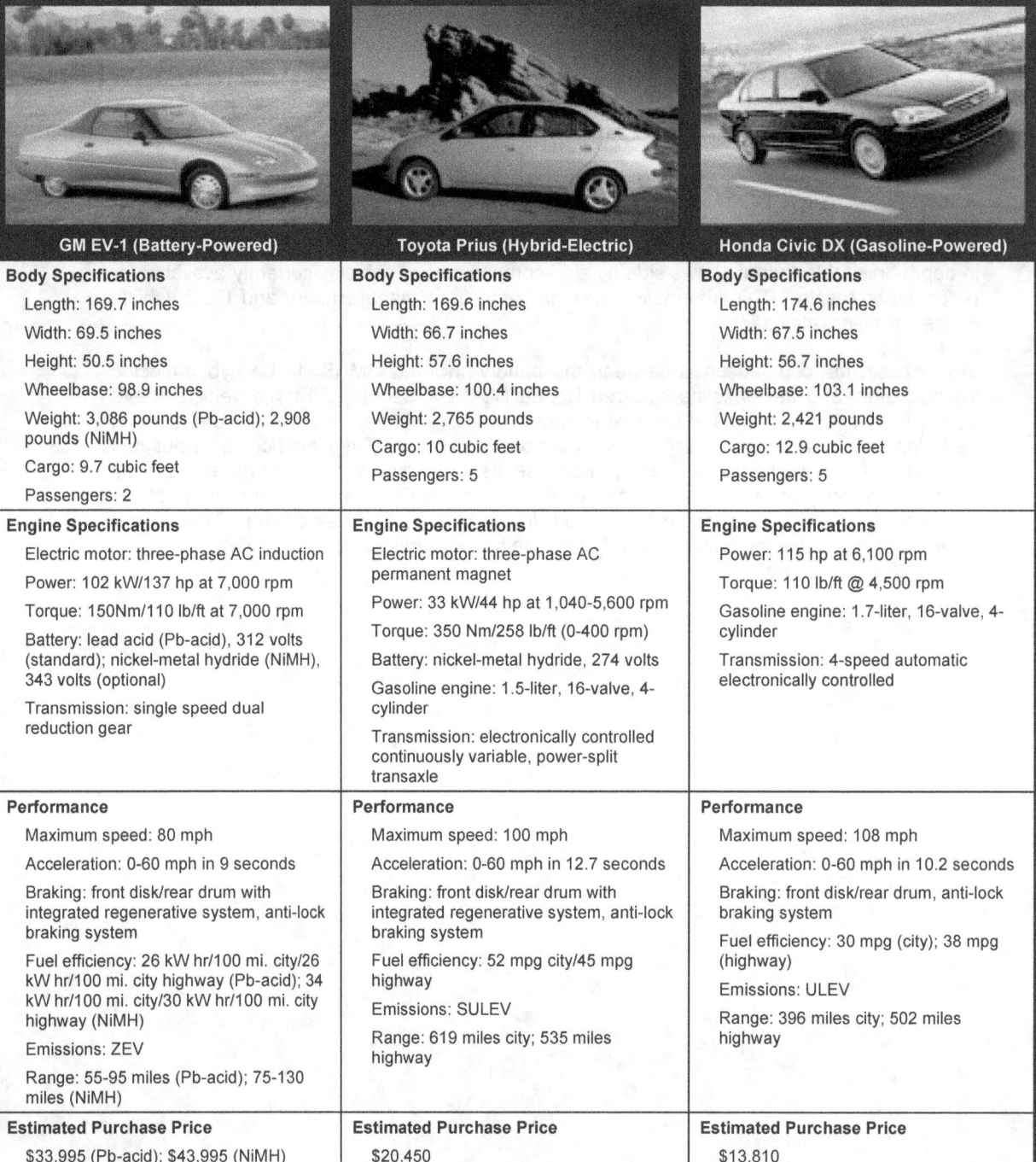

GM EV-1 (Battery-Powered)	Toyota Prius (Hybrid-Electric)	Honda Civic DX (Gasoline-Powered)
Body Specifications	**Body Specifications**	**Body Specifications**
Length: 169.7 inches	Length: 169.6 inches	Length: 174.6 inches
Width: 69.5 inches	Width: 66.7 inches	Width: 67.5 inches
Height: 50.5 inches	Height: 57.6 inches	Height: 56.7 inches
Wheelbase: 98.9 inches	Wheelbase: 100.4 inches	Wheelbase: 103.1 inches
Weight: 3,086 pounds (Pb-acid); 2,908 pounds (NiMH)	Weight: 2,765 pounds	Weight: 2,421 pounds
Cargo: 9.7 cubic feet	Cargo: 10 cubic feet	Cargo: 12.9 cubic feet
Passengers: 2	Passengers: 5	Passengers: 5
Engine Specifications	**Engine Specifications**	**Engine Specifications**
Electric motor: three-phase AC induction	Electric motor: three-phase AC permanent magnet	Power: 115 hp at 6,100 rpm
Power: 102 kW/137 hp at 7,000 rpm	Power: 33 kW/44 hp at 1,040-5,600 rpm	Torque: 110 lb/ft @ 4,500 rpm
Torque: 150Nm/110 lb/ft at 7,000 rpm	Torque: 350 Nm/258 lb/ft (0-400 rpm)	Gasoline engine: 1.7-liter, 16-valve, 4-cylinder
Battery: lead acid (Pb-acid), 312 volts (standard); nickel-metal hydride (NiMH), 343 volts (optional)	Battery: nickel-metal hydride, 274 volts	Transmission: 4-speed automatic electronically controlled
Transmission: single speed dual reduction gear	Gasoline engine: 1.5-liter, 16-valve, 4-cylinder	
	Transmission: electronically controlled continuously variable, power-split transaxle	
Performance	**Performance**	**Performance**
Maximum speed: 80 mph	Maximum speed: 100 mph	Maximum speed: 108 mph
Acceleration: 0-60 mph in 9 seconds	Acceleration: 0-60 mph in 12.7 seconds	Acceleration: 0-60 mph in 10.2 seconds
Braking: front disk/rear drum with integrated regenerative system, anti-lock braking system	Braking: front disk/rear drum with integrated regenerative system, anti-lock braking system	Braking: front disk/rear drum, anti-lock braking system
Fuel efficiency: 26 kW hr/100 mi. city/26 kW hr/100 mi. city highway (Pb-acid); 34 kW hr/100 mi. city/30 kW hr/100 mi. city highway (NiMH)	Fuel efficiency: 52 mpg city/45 mpg highway	Fuel efficiency: 30 mpg (city); 38 mpg (highway)
Emissions: ZEV	Emissions: SULEV	Emissions: ULEV
Range: 55-95 miles (Pb-acid); 75-130 miles (NiMH)	Range: 619 miles city; 535 miles highway	Range: 396 miles city; 502 miles highway
Estimated Purchase Price	**Estimated Purchase Price**	**Estimated Purchase Price**
$33,995 (Pb-acid); $43,995 (NiMH)	$20,450	$13,810

A1 Comparison of Vehicles

A2   Lifecycle Ownership Cost Analysis

A study by Argonne National Laboratory analyzes and compares lifecycle ownership costs of EVs and conventional vehicles (CV).[129] Lifecycle ownership cost includes amortized purchase price, operating costs, and other incidental costs. Operating costs include energy, battery costs, and maintenance costs. These costs are then converted to per-mile costs, and are comparable with similar costs for a CV. The study considered two cases: one at the time of EV introduction, and another at the time EVs are sold in high volume (more than 100,000 units).

In computing the lifecycle costs, several assumptions were made. Vehicle-ownership-related costs such as registration, insurance, and property taxes, are excluded, assuming that these items would be the same for both types of vehicles. Similar annual and lifetime usage of the vehicles at 130,000 miles over a span of 12 years with similar scrappage value at end of their useful lives was assumed. The EV characteristics required for this study included an acceleration of zero to 60 mph in 10-12 seconds, and a battery pack with an energy content of at least 15 kWh (nickel-metal hydride or lead-acid).

The EV operating costs included initial and replacement battery packs; routine maintenance costs 20% lower than CVs; and tire cost 11% higher than CVs. The high volume test case accounted for improvements made to EVs, including a 3% reduction in vehicle mass, an 8% power train efficiency improvement, and a 4% discount to the price of replacement battery packs (excluding inflation). Other considerations should be taken into account when comparing CVs and EVs; for example, the dissimilarities in refueling infrastructure and the range between refueling stations. Some researchers have analyzed the perceived negative value held by consumers and translated this utility into a cost equivalent.[130]

---

[129] R.M. Cuenca, L.L. Gaines, and A.D. Vyas, "Evaluation of Electric Vehicle Production and Operating Costs," Argonne National Laboratory, November 1999.

[130] For further study, see Bunch, D.S., et al., 1991, "Demand for Clean-Fuel Personal Vehicles in California: A Discrete-Choice Stated Preference Survey," Institute of Transportation Studies, University of California at Irvine, Report UCT-ITS-WP-91-8, Irvine, CA; and Tompkins, M., et al., 1998, "Determinants of Alternative Fuel Vehicle Choice in the Continental United States," Transportation Research Record No. 1641, Transportation Research Board, Washington, DC.

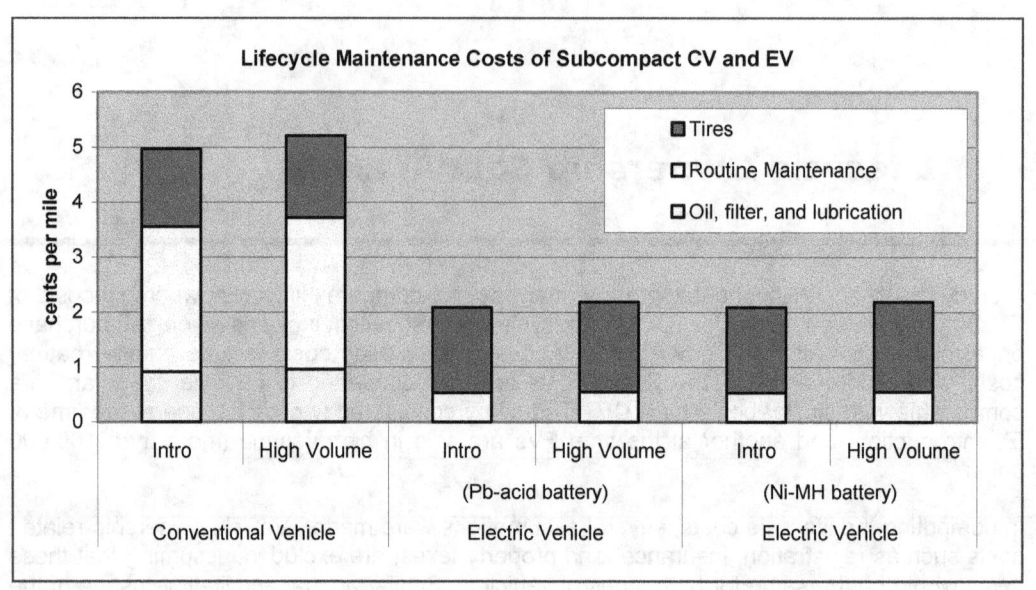

The lifecycle energy costs uses energy use rates measured in miles per gallon for CVs and watt hours per miles for EVs. CVs have an on-road fuel economy factor of 0.8; it is 1 for EVs. The energy price for CVs is measures in dollars per gallon, and dollars per kilowatt hour for EVs. EVs' energy costs were measured with a consumption rate of 111 Wh/km per metric ton vehicle mass and used off-peak electricity rates.

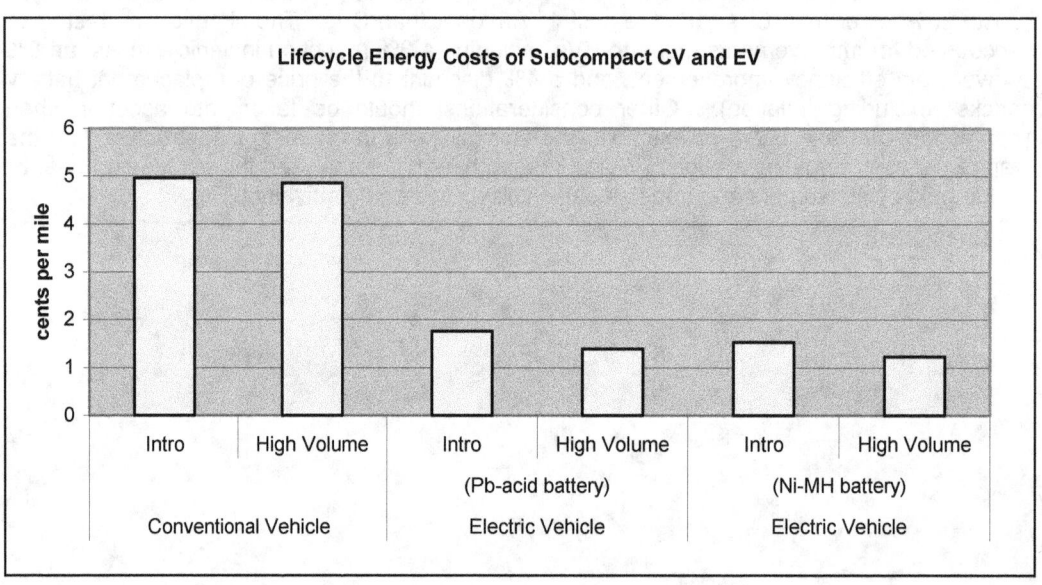

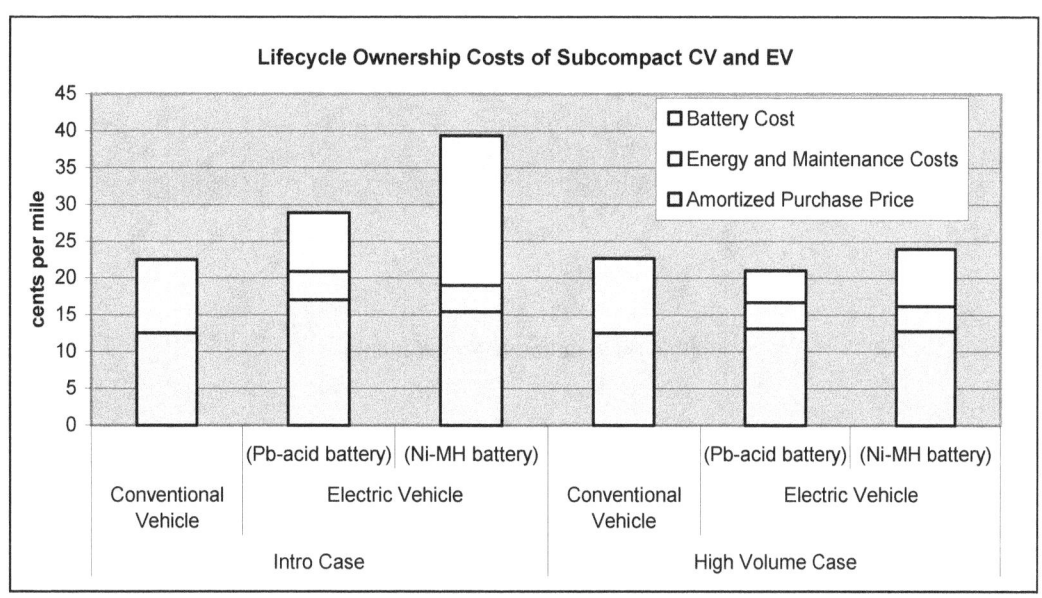

In lifetime ownership costs, the CV has a substantial advantage over the EV in the Intro case. The energy and maintenance costs, in both cases, are lower for EVs, but the lifetime cost advantage is lost when the price of the battery is added. Given the study's assumptions about long-term cost reductions potential for the electric drive, and advanced battery performance, the EV could be comparable to the CV.

[this page deliberately left blank]

A3 U.S. State Registries for Reporting of Greenhouse Gases and State Legislation/Policies to Promote GHG Emission Reductions

GHG registries are designed as tools to help entities (e.g. companies, households, individuals) that are interested in quantifying and recording their efforts to reduce GHG emissions. Registries also publicly record their progress in reducing emissions and provide public recognition of a company's accomplishments. Registries can also help raise awareness of climate change, promote sharing of lessons learned and success stories, and publicize low-cost mitigation opportunities.

This appendix provides a summary of 18 recent states', counties', and regional associations' actions and legislation towards developing GHG registries. Up until recently, a lack of federal leadership on the role of GHG registries (at the state or national level) had prompted many states and regions to begin to take action and resulted in an emerging patchwork of differing requirements and approaches. However, on February 14, 2002, President Bush introduced the Administration's official policy on climate change:

> Our immediate goal is to reduce America's GHG emissions relative to the size of our economy... Our government will also move forward immediately to create world-class standards for measuring and registering emission reductions. And we will give transferable credits to companies that can show real emission reductions.[131]

The President's Global Climate Change Policy Book specifically addresses local and national GHG registries:

> The President directed the Secretary of Energy, in consultation with the Secretary of Commerce, the Secretary of Agriculture, and the Administrator of the EPA, to propose improvements to the current voluntary emission reduction registration program under section 1605(b) of the 1992 Energy Policy Act within 120 days... A number of proposals to reform the existing registry—or create a new registry—have appeared in energy and/or climate policy bills introduced in the past year. The Administration will fully explore the extent to which the existing authority under the Energy Policy Act is adequate to achieve these reforms.[132]

Many states are reacting to the Administration's policy and are eager to respond to the recommendations expected after 120 days. Project developers should keep abreast to current events with respect to these emerging registry programs.

---

[131] White House Office of the Press Secretary, "President Announces Clear Skies & Global Climate Change Initiatives," February 14, 2002, http://www.whitehouse.gov/news/releases/2002/02/20020214-5.html.
[132] White House, Global Climate Change Policy Book, February 2002, http://www.whitehouse.gov/news/releases/2002/02/climatechange.html.

[this page deliberately left blank]

Table A3   U.S. State Registries for Reporting of Greenhouse Gases and State Legislation/Policies to Promote GHG Emission Reductions

Region/State/City	Directive	Date	Objective	Contact
California	Senate Bill 1771	Signed September 30, 2000	Specified the creation of the non-profit organization, the California Climate Action Registry (California Registry). The California Registry will help various California entities' to establish GHG emissions baselines. Also, the California Registry will enable participating entities to voluntarily record their annual GHG emissions inventories. In turn, the State of California will use its best efforts to ensure that organizations that voluntarily inventory their emissions receive appropriate consideration under any future international, federal, or state regulatory regimes relating to GHG emissions.[133]	For more information, contact California Energy Commission Climate Change Program Manager. Pierre duVair, Ph.D., tel. 916-653-8685 or email pduvair@energy.state.ca.us.
	Senate Bill 527	Signed October 13, 2001	This bill requires the California Energy Commission to provide guidance to the California Registry on a number of issues, such as, developing GHG emissions protocols, qualifying third-party organizations to provide technical assistance, and qualifying third-party organizations to provide certification of emissions baselines and inventories.[134]	For more information, contact California Energy Commission Climate Change Program Manager. Pierre duVair, Ph.D., tel. 916-653-8685 or email pduvair@energy.state.ca.us.
Illinois	Senate Bill 372[135]	Signed September, 2001[136]	This bill requires the Illinois EPA to establish an interstate nitrogen oxide trading program and issue findings that address the need to control or reduce emissions from fossil fuel-fired electric generating plants. The findings are to address the establishment of a banking system, consistent with DOE's Voluntary Reporting of Greenhouse Gases Program for certifying credits for voluntary offsets of emissions of greenhouse gases, or reductions of GHGs.[137]	For more information, contact Steven King at the Illinois Environmental Protection Agency, tel. 217-524-4792, or email steven.king@epa.state.il.us.
Maine	Legislative Document 87[138]	Passed April 6, 2001	This requires the Department of Environmental Protection to develop rules to create a voluntary registry of GHG emissions. The rules must provide for the collection of data on the origin of the carbon emissions as either fossil fuel or renewable resources, and the collection of data on production activity to allow the tracking of future emission trends.	For more information, the bill sponsor, Re. Robert Daigle, tel. (800) 423-2900, or e-mail rdaigle@gwi.net; and the contact the Maine Department of Environmental Protection, tel. 800-452-1942.

133 California Energy Commission, Global Climate Change & California, http://www.energy.ca.gov/global_climate_change/index.html.
134 California Energy Commission, Global Climate Change & California, http://www.energy.ca.gov/global_climate_change/index.html.
135 Full text of Illinois Senate Bill 372 can be read at website, http://www.legis.state.il.us/scripts/imstran.exe?LIBSINCWSB372.
136 Illinois State Senate Democrat News, September 9, 2001, http://www.senatedem.state.il.us/senatenews/news.shtml.
137 U.S. Environmental Protection Agency, Legislative Initiatives, http://yosemite.epa.gov/globalwarming/ghg.nsf/actions/LegislativeInitiatives.
138 Full text of Maine Legislative Document 87 can be read at website, http://janus.state.me.us/legis/bills/.

A3 State Legislation, Policies, and Registries

Table A3   U.S. State Registries for Reporting of Greenhouse Gases and State Legislation/Policies to Promote GHG Emission Reductions

Region/State/City	Directive	Date	Objective	Contact
Maryland	Executive Order 01.01.2001.02, "Sustaining Maryland's Future with Clean Power, Green Buildings and Energy Efficiency"	Signed March 13, 2001	It states, "the [established] [Maryland Green Buildings] Council shall develop a comprehensive set of initiatives known as the 'Maryland Greenhouse Gas Reduction Action Plan;' and The Council shall report annually to the Governor and to the General Assembly on the efforts of State agencies in the implementation of... the Greenhouse Gas Reduction Plan, and other energy efficiency, energy production and sustainability issues or policies the Council may have considered. [139] A November 2001 report by the Council stated, "Goals [for GHG reductions in Maryland] will be set for both the State facilities and operations as well as statewide reduction goals to be achieved through voluntary initiatives, policies, and programs."[140]	For more information on GHG reduction plans in Maryland, contact the Maryland Green Buildings Council at http://www.dgs.state.md.us/GreenBuildings/default.htm, Gerri Nicholson, Greenhouse Gas Reduction Plan, tel. 410-260-7207, or e-mail gnicholson@energy.state.md.us.
Massachusetts	Department of Environmental Protection Regulation 310 CMR 7.29	Issued April 23, 2001	This requires the six highest-polluting power plants in Massachusetts to meet overall emission limits for nitrogen oxide and sulfur dioxide by October 1, 2004 and begin immediate monitoring and reporting of mercury emissions. For the six affected plants, the rule caps total carbon dioxide emissions and creates an emission standard of 1,800 lbs. of carbon dioxide per megawatt-hour (a reduction of 10 percent below the current average emissions rate). The carbon dioxide limits must be met by October 1, 2006 or October 1, 2008 for plant retrofit or replacement. Plant operators may meet the standard either by increasing efficiency at the plant, or by purchasing credits from other reduction programs approved by the Department of Environmental Protection.[141]	For more information, contact the Department of Environmental Protection InfoLine, tel. 617-338-2255 or 800-462-0444, or email dep.infoline@state.ma.us; or for Emissions Trading, contact Bill Lamkin, tel. 978-661-7657 or email Bill.Lamkin@state.ma.us; or for the Air Program Planning Unit that covers these regulations, see Nancy Seidman, tel. 617-556-1020, or email Nancy.Seidman@state.ma.us.
Michigan	Senate Bill 693	Introduced October 2001	This bill to amend the 1994 Natural Resources and Environmental Protection Act calls for declining caps in nitrogen oxides, sulfur dioxide, carbon dioxide, and mercury by 2007.[142] The bill has been referred to the Committee on Natural Resources and Environmental Affairs.[143]	For more information, contact the bill sponsor, Senator Alma Wheeler Smith, tel. 800-344-2562 or 517-373-2406 or email SenASmith@senate.state.mi.us.

139 Full text of the State of Maryland, Executive Order 01.01.2001.02, can be read at website, http://www.gov.state.md.us/gov/execords/2001/html/0002eo.html.
140 Maryland Green Buildings Council, "2001 Green Buildings Council Report," November 2001, pg. 30, http://www.dgs.state.md.us/GreenBuildings/Documents/FullReport.pdf.
141 U.S. Environmental Protection Agency, Legislative Initiatives, http://yosemite.epa.gov/globalwarming/ghg.nsf/actions/LegislativeInitiatives.
142 Jones, Brian M., "Emerging State and Regional GHG Emission Trading Drivers," presented at the Electric Utilities Environmental Conference, Tuscan, Arizona, January 2002.
143 Michigan State Legislature, Senate Bill 0693, http://www.mileg.org.

A3 State Legislation, Policies, and Registries

Table A3   U.S. State Registries for Reporting of Greenhouse Gases and State Legislation/Policies to Promote GHG Emission Reductions

Region/State/City	Directive	Date	Objective	Contact
New England Governors/ Eastern Canadian Premiers	Climate Action Plan[144]	Signed August, 2001	The climate change action plan defines incremental goals for the coalition: in the short-term, reduce regional GHG emissions to 1990 emissions by 2010; for the mid-term, reduce regional GHG emissions by at least 10 percent below 1990 emissions by 2020, and establish an iterative five-year process, beginning in 2005, to adjust existing goals, if necessary, and set future emissions reduction goals; and for the long-term, reduce regional GHG emissions sufficiently to eliminate any dangerous threat to climate; current science suggests this will require reductions of 75 percent-85 percent below current levels. The action plan calls for the creation of a regional emissions registry and the exploration of a trading mechanism.	For more information, contact the New England Secretariat, New England Governors' Conference Inc., tel. 617-423-6900 or email negc@tiac.net.
New Hampshire	House Bill 284, "Clean Power Act"	Approved January 2, 2002	This four-pollutant bill is the first in the nation to include carbon dioxide.[145] Emission reduction requirements include 75% of sulfur dioxide by 2006; 70% of nitrogen oxide by 2006; 3% of carbon dioxide by 2006 (1990 levels); and mercury levels are still to be determined by 2004.[146]	For more information, contact New Hampshire Office of the Governor, tel. 603-271-2121.
	Senate Bill 159	Approved July 6, 1999	This bill established a registry for voluntary GHG emission reductions to create an incentive for voluntary emission reductions.[147] Implementation rules were adopted on February 23, 2001.	For more information, contact Joanna Morin, Department of Environmental Science, tel. 800-498-6868 or 603-271-1370, or email jmorin@desstate.nh.us.
New Jersey	N.J.A.C 7:27-30.2 and 30.5	Adopted April 17, 2000	The New Jersey Department of Environmental Protection adopted new rules to add provisions to the Open Market Emissions Trading Program for the generation and banking of GHG credits.[148] The GHGs included are: carbon dioxide; methane; nitrous oxide; certain hydro fluorocarbons, per fluorocarbons; and sulfur hexafluoride. The Program was established to provide incentives for voluntary reduction of air contaminant emissions and also provide an alternative means for regulated entities to achieve compliance with air pollution control obligations in a more cost-effective manor. Read the draft	For more information, contact the New Jersey Department of Environmental Protection, Air Quality Management Bureau of Regulatory Development, tel. 609-777-1345 or email aqrdweb@dep.state.nj.us.

---

[144] Full text of the New England Governors/ Eastern Canadian Premiers Climate Change Action Plan can be read at website http://www.cmp.ca/CCAPe.pdf.
[145] New Hampshire, Office of the Governor, Press Releases, "Governor Shaheen Hails House Passage of Clean Power Act," http://www.state.nh.us/governor/media/010202clean.html.
[146] Jones, Brian M., "Emerging State and Regional GHG Emission Trading Drivers," presented at the Electric Utilities Environmental Conference, Tuscan, Arizona, January 2002.
[147] New Hampshire, Senate Bill 0159, http://www.gencourt.state.nh.us/legislation/1999/sb0159.
[148] New Jersey, Department of Environmental Protection, Air Quality Permitting, Air Quality Management, Air And Environmental Quality Enforcement, "Open Market Emissions Trading Rule," http://www.state.nj.us/dep/aqm/ometp2ad.htm.

A3 State Legislation, Policies, and Registries

Table A3   U.S. State Registries for Reporting of Greenhouse Gases and State Legislation/Policies to Promote GHG Emission Reductions

Region/State/City	Directive	Date	Objective	Contact
			guidance on the preparation of quantification protocols at www.state.nj.us/dep/aqm/omet/.	
New York State	Greenhouse Gas Task Force	Created June, 2001	Governor Pataki set up a Greenhouse Gas Task Force in to come up with policy recommendations on climate change. [149] Preliminary recommendations for actions and policies from the Task Force's Working Groups include establishing a statewide target for GHG emission reductions relative to 1990 levels, and establishing a greenhouse gas registry to document baseline emissions and voluntary emissions reductions for participating customers. The Task Force plans a Final Report to be complete by March 2002. [150]	For more information on the Greenhouse Gas Task Force, check the Governor's website at www.state.ny.us/governor.
	Assembly Bill 5577	Introduced February 27, 2001	This bill provides for regulation of emissions of nitrogen oxide, sulfur dioxide and carbon dioxide. This bill passed the Assembly on March 25, 2002, and has been referred in the Senate to the Environmental Conservation Committee. [151]	For more information, contact the bill sponsor, Richard Brodsky, tel. 518-455-5753 or 914-345-0432, or email brodskr@assembly.state.ny.us.
New York City	New York City Council Bill No. 30 [152]	Re-introduced January 30, 2002	New York City Council member Peter Vallone Jr. reintroduced a bill that would require the city's power plants to reduce carbon dioxide emissions or face stiff fines. If passed, the bill would reduce carbon dioxide emissions by as much as 20 percent within five years of enactment. Under the terms of the legislation, owners of power plants that produce at least 25 megawatts of electricity for sale would be required to pay high fines for generators that emit levels of carbon dioxide that exceed those established by an independent board. [153]	For more information, contact Council member Peter Vallone Jr. tel. 718 274-4500 or 212-788-6963, or email vallonejr@council.nyc.ny.us.
Suffolk County	Carbon Dioxide Law	Passed July 24, 2001	Suffolk County became the first county to pass a resolution limiting carbon dioxide emissions. The resolution seeks to encourage efficiency in existing power plants and future facilities by setting allowable rates for carbon dioxide emissions and penalties for exceeding those limits. Under the law taking affect March 1, 2002 any power plant in the county that generates over 1,800 pounds of carbon dioxide emissions per Megawatt/hour	For more information in Suffolk County, contact Suffolk County Executive's Office, tel. 631-853-4000.

---

[149] Press Release, Office of the Governor of New York State, June 10, 2001, http://www.state.ny.us/governor/press/year01/june10_01.html.
[150] New York State, Draft State Energy Plan, December 2001, http://www.nyserda.org/draftsepsec2.pdf.
[151] New York State Assembly, Bill 5577, http://www.assembly.state.ny.us/leg/?bn=A.5577.
[152] Full text of New York City Council Bill Int. No. 30, can be read at website, http://www.council.nyc.ny.us/pdf_files/bills/int0030-2002.htm.
[153] Forbes, "NYC Council Seeks Cut In Power Plant CO2 Emissions," January 30, 2002, www.forbes.com/newswire/2002/01/30/rtr498771.html.

A3 State Legislation, Policies, and Registries

Table A3   U.S. State Registries for Reporting of Greenhouse Gases and State Legislation/Policies to Promote GHG Emission Reductions

Region/State/City	Directive	Date	Objective	Contact
			would be fined two dollars for every ton above the limit. An additional $1 per excess ton would be charged in each consecutive year. The bill contains several alternatives to paying fines including buying emission credits through nationally recognized carbon dioxide trading markets, investing in alternative energy sources or donating penalties to community environmental groups.[154]	
Nassau County	Carbon Dioxide Proposal	N/A	A carbon dioxide emissions rate of 1,800 lbs/MWh is proposed, along with an allowable County-wide emission rate reduction of 1% for every 100 MW of electric generating capacity installed within the County until the emission rate has been reduced by 20%. Emissions trading would be allowed for compliance. Penalties of $2/ton in the first year and $1/ton each consecutive year would be assessed if the EGU fails to comply.[155]	For more information, contact Nassau County Government at tel. 516-571-3000.
North Carolina	Senate Bill 1078 (House Bill 1015), "Clean Smokestack Bill"	Passed the Senate on April 23, 2001	This bill would reduce emissions of nitrogen oxides by 78% by 2009 and sulfur dioxide by 73% by 2013. The bill also directs the North Carolina Division of Air Quality (DAQ) to study the issues for standards of reductions of mercury and carbon dioxide. DAQ is also to develop and adopt a program of incentives to promote voluntary reductions of emissions including, emissions banking and trading and credit for voluntary early action. This bill was sent by the Senate to the House where is has been referred the Committee on Public Utilities[156]	For more information, contact the North Carolina Division of Air Quality at tel. 919-733-3340, or for Climate Change and Greenhouse Gases, contact Russell Hageman at tel. 919-733-1490 or email Russell.Hageman@ncmail.net, Jill Vitas at tel. 919-715-8666 or email Jill.Vitas@ncmail.net.
Oregon	House Bill 3283	Signed June 26, 1997	This bill established a carbon dioxide standard requiring new utilities to emit 17% less than most energy efficient plant available.[157] The bill capped carbon dioxide emissions at 0.7 pounds of carbon dioxide per kilowatt-hour for base-load natural gas-fired power plants; in 1999 the cap was lowered to 0.675 pounds per kilowatt-hour. New energy facilities built in the state must avoid, sequester, or pay a per-ton of carbon dioxide offset into the Oregon Climate Trust.[158] The nonprofit Oregon Climate Trust accepts mitigation funds from energy facilities for displacing	For more information on purchasing carbon dioxide offsets in Oregon, or applying for project funding for new carbon dioxide mitigation projects, contact Mike Burnett, Executive Director, tel. 503-238-1915 or email: info@climatetrust.org and see website www.climatetrust.org.

[154] Suffolk County, Press Release, "Suffolk Becomes First County to Limit CO2 Emissions," July 24, 2001, http://www.co.suffolk.ny.us/exec/press/2001/emissions.html.
[155] Jones, Brian M., "Emerging State and Regional GHG Emission Trading Drivers," presented at the Electric Utilities Environmental Conference, Tuscan, Arizona, January 2002.
[156] North Carolina General Assembly, Senate Bill 1078 (also called House Bill 1015), http://www.ncga.state.nc.us/gascripts/billnumber/billnumber.pl?Session=2001&BillID=S1078.
[157] Full text of Oregon House Bill 3283 can be read at website, http://www.leg.state.or.us/97reg/measures/hb3200.dir/hb3283.int.html.
[158] U.S. Environmental Protection Agency, Legislative Initiatives, http://yosemite.epa.gov/globalwarming/ghg.nsf/actions/LegislativeInitiatives.

A3 State Legislation, Policies, and Registries

Table A3   U.S. State Registries for Reporting of Greenhouse Gases and State Legislation/Policies to Promote GHG Emission Reductions

Region/State/City	Directive	Date	Objective	Contact
			their unmet emissions requirements, and in turn must use the funds to carry out projects that avoid, sequester, or displace the carbon dioxide. In January 2001, the Climate Trust released a request for proposals (RFP) to fund $5.5 million in carbon dioxide mitigation projects.	
Texas	Texas Natural Resources Conservation Commission, report on greenhouse gases and recommendations from the Executive Director	Presented January 18, 2002	In August 2000, the Texas Natural Resources Conservation Commission (TNRCC) issued a decision instructing the agency's Executive Director to prepare a report on GHGs. The draft report and recommendations from the Executive Director were presented to TNRCC commissioners at a public work session on January 18, 2002. The recommendations included, "Develop and maintain a voluntary registry for reporting GHG emission reductions resulting from specific emission reduction or sequestration projects and energy efficiency improvements within Texas.* [*The Chairman directed staff, before executing this recommendation, to evaluate the DOE 1605(b) voluntary greenhouse gas registry program, as is or with some changes, as a possible element of a Texas GHG registry which avoids duplicative reporting.]"[159]	For current information on the status of TNRCC efforts, contact the Office of Environmental Policy, Analysis and Assessment by calling 512-239-4900, or email policy@tnrcc.state.tx.us.
Washington	Senate Bill 5674	Passed the House on March 13, 2001	Senate Bill 5674, was passed by the House on March 13, 2001 and was referred on motion to the Environment, Energy & Water Committee.[160] This bill authorizes the establishment of an independent, nonprofit organization known as the Washington Climate Center to serve as a central clearinghouse for all climate change activities in the state. The Climate Center's activities include determining current and projected GHG emissions in the state, and studying and recommending the most cost-effective methods for reducing all net GHG emissions.[161]	For more information, contact one of the Bill sponsors, Sen. Ken Jacobsen at tel. 360-86-7690 or email jacobsen_ke@leg.wa.gov; Sen. Margarita Prentice at tel. 360-786-7616 or email prentice_ma@leg.wa.gov; Sen. Karen Fraser at tel. 360-786-7642 or email fraser_ka@leg.wa.gov; Sen. Jeanne Kohl-Welles at tel. 360-786-7670 or email kohl_je@leg.wa.gov, or former Sen. Dow Constantine at tel. 206-296-1008 or email dow.constantine@metrokc.gov.

[159] Texas Natural Resource Conservation Commission, Office of Environmental Policy, Analysis and Assessment, "Overview and Recommendations Identified by A Report to the Commission on Greenhouse Gases," February 8, 2002, http://www.tnrcc.state.tx.us/oprd/sips/greenhouse/.
[160] Full text of Washington Senate Bill 5674 can be read at website, http://www.leg.wa.gov/wsladm/billinfo/dspBillSummary.cfm?billnumber=5674.
[161] U.S. Environmental Protection Agency, Legislative Initiatives, http://yosemite.epa.gov/globalwarming/ghg.nsf/actions/LegislativeInitiatives.

A3 State Legislation, Policies, and Registries

Table A3   U.S. State Registries for Reporting of Greenhouse Gases and State Legislation/Policies to Promote GHG Emission Reductions

Region/State/City	Directive	Date	Objective	Contact
Wisconsin	Assembly Bill 627	February 8, 2000	This bill requires the Department of Natural Resources to establish and operate a system for registering reductions in emissions of GHGs if the reductions are made before they are required by law. The bill authorizes the Department of Natural Resources to establish systems for registering reductions in fine particulate matter, mercury and other air contaminants.[162]	For more information, contact the Wisconsin Voluntary Emissions Reductions Registry Advisory Committee at http://www.dnr.state.wi.us/org/aw/air/hot/climchgcoml/.
	Department of Natural Resources Rule NR 437	N/A	A rule (NR 437) is proposed to establish voluntary emissions reduction registries for GHGs, as well as for mercury, fine particulate matter and other contaminants that cause air pollution. The rule represents a new Department of Natural Resources policy to systematically record and track voluntary emission reductions by industries, electric utility companies, agricultural and forestry interests, and transportation and energy efficiency interests. NR 437 establishes the rules and procedures under which the new registry will operate. The rule also identifies the sources that are eligible to register reductions for GHGs like carbon dioxide, methane, nitrous oxide, hydrofluorocarbons, perfluorcarbons and sulfur hexafluoride, as well as for nitrogen oxides, sulfur dioxide, volatile organic compounds, carbon monoxide, mercury, lead and fine particulate matter.	For more information, contact Eric Mosher at tel. 608-266-3010, or e-mail moshee@dnr.state.wi.us.

---

[162] Full text of Wisconsin Assembly Bill 627 can be read at website, http://www.legis.state.wi.us/1999/data/AB627.pdf.

A3 State Legislation, Policies, and Registries

A4 Electric and Hybrid-Electric Vehicle Year 2000 Projects Reported to the U.S. Voluntary Reporting of Greenhouse Gases Program (1605(b))

Table A4 Electric and Hybrid-Electric Vehicle Year 2000 Projects Reported to the U.S. Voluntary Reporting of Greenhouse Gases Program

Reporting Entity	Project Name	Project Size *	Reported CO_2 Equivalent Reduction in 2000 (metric tons)
Allegheny Energy, Inc.			
Carryall Vehicle Program		15 Vehicles	Direct: 15.17 Indirect: 796.51
Project Description	The Pleasants & Willow Island Power Stations are adjacent coal-fired power generation facilities along the Ohio River near Parkersburg, West Virginia. Plant personnel had used a fleet of pick-up trucks and vans to perform various duties associated with the operation and maintenance of the plant. The number of vehicles in the fleet at the plant totaled 27 at one time. Beginning in 1992, the plant began to evaluate the use of Carryall vehicles to replace the pick-up trucks. These utility vehicles, which are similar to golf carts, were acquired because of the potential to reduce costs to operate and maintain the vehicle fleet through lower purchase price, reduced fuel consumption, and reduced maintenance costs. Both gasoline-powered and electric-powered versions of the Carryall vehicle have been acquired. After a period of demonstration and evaluation, the decision was made in 1996 to retire a portion of the vehicle fleet and to use the Carryall vehicles as the primary vehicles for in-plant transportation. By 1997, the power station complex has added a total of 15 Carryall vehicles to its fleet. There are 12 gasoline-powered vehicles and 3 electric-powered vehicles.		
Estimation Method	Only estimates of changes in CO_2 emissions have been included in this report. The 15 pick-up trucks and other vehicles replaced averaged about 350 gallons/truck of fuel consumption annually. The 12 gasoline-powered Carryalls consume about 85 gallons of fuel each on an annual basis. The emissions factor of 19.641 lbs.CO2/gal for each gallon of fuel consumed was applied to the difference in fuel consumption to determine the reduction in CO_2 emissions.		

A4 Project Reports to 1605(b)

A4 Project Reports to 1605(b)

The CO_2 emitted from the electricity generated to charge the electric Carryalls was subtracted from the difference mentioned above to determine the net reduction in CO_2. The electric-powered Carryalls are charged each night for about 10 hours. $CO2$ emissions for the electricity used are about 1.05 tons/MWh from the Pleasants/Willow Island Power Station complex. The MWh consumption was determined based on information supplied by the vendor. From this information the total CO2 emissions from battery charging can be determined.

The vehicle replacement program was considered to be fully implemented in October 1996 when the majority of vehicles were removed from service, so 1996 figures represent 25 percent of annual emissions/reductions. Beginning in 1997, data is reported on an annual basis.

DTE Energy/Detroit Edison		
Electric Vehicle Demonstration Project	10 vehicles (1994-96); 27 vehicles (1997-98)	N/A
Project Description	Detroit Edison has completed a 30-month Ford Ecostar demonstration program. There were 10 Ecostar electric vans being used in various Company fleet applications. The Ecostars were also displayed in public automotive events and alternative fuel vehicle industry conferences and events.	
	Detroit Edison continues to promote educational efforts to enhance understanding and use of electric vehicles. Sponsoring the 1996 Future Car Challenge, the Hybrid Electric Vehicle Challenges (1993-1995) and past SAE Micro Electric Vehicle Challenges are part of this effort. The Company also sponsored the 1996 EEI Fleet Managers Electric Vehicle Conference and participated in planning and execution of the EV Ready Market Launch Workshop in Detroit.	
	The 10 Ford Ecostar vans were retired in 1997, beginning in June, September and November. Since the vehicles were only utilized for a relatively short time in 1997, the CO_2 reductions are expected to be greater during 1998. Detroit Edison purchased 27 GM S10 electric pickup trucks. The pickups were delivered in June, September, and November.	
	The fleet of 27 GMC S10 electric pickup trucks were used throughout 1998 but did not result in the anticipated reduction in CO_2 emissions because of lower efficiency of these units.	
	In 1999, operation/use of our fleet of electric vehicles was transferred to a local hospital. This precluded accurate record-keeping on electricity used, miles driven and mode of use (e.g., driving vs. standing/idling). Therefore, no report is submitted for 1999 or subsequent years.	
	Detroit Edison's formal involvement with the electric vehicle assessment program concluded at the end of calendar 1998. No further reports will be submitted on this project.	
Estimation Method	The electric vehicles replaced various types of internal combustion engine vehicles which had had fuel economies of 4 to 20 miles per gallon, depending on application.	
	Fuel savings (gallons gasoline) were estimated according to the fuel economies (miles per gallon) of the vehicles replaced and the actual miles driven by the electric vehicles.	

The associated CO_2 emissions displaced were calculated by multiplying the gallons of gasoline saved by 19.564 lb CO2/gal. (Appendix B, Form 1605 Instructions) lbs CO_2 was then converted to short tons.

The electricity consumption from vehicle charging was obtained by metering on-site.

Emissions from vehicle charging were calculated by multiplying the MWh consumption by Detroit Edison's annual fossil CO2 emission rate.

Annual fossil generation CO_2 emission rates:

1994 1.052 short ton/MWh
1995 1.061 short ton/MWh
1996 1.073 short ton/MWh
1997 1.069 short ton/MWh
1998 1.058 short ton/MWh

Note: Electricity consumption for 1997 was estimated by: MWh electricity consumed = Total miles driven / (3 miles per kWh * 1000 kWh per MWh).

CO_2 Emissions (short tons)

Year	Decrease due to gas not burned	Increase due to electrical generation	Net Reduction
1994	18.2	10.4	7.8
1995	55.3	47.8	7.5
1996	65.4	62.3	3.1
1997	42.4	22.9	19.5
1998	48.4	37.5	10.9

GPU, Inc.

Electric Vehicles and Employee Trip Reduction Program	1,049,106 VMT eliminated/displaced by EV over 6 years	Direct: 12.28
	36,012 VMT in 2000	Indirect: 3.06
Project Description	GPU has undertaken various programs in New Jersey to: encourage ridematching; sponsor vanpooling; and facilitate carpooling as well as the use of mass transit services. An official program was prepared for submission to the NJDOT.	
	GPU has also undertaken voluntary measures to explore the use of alternatively fueled vehicles for different company functions, including meter reading and shuttle services. Three cars were put into service during 1995, collectively driving 6800 miles.	
	All recorded Vehicle Miles Traveled (VMT) is mileage by Electric Vehicle that displaced mileage by gasoline fueled vehicles.	

	Note: The ridesharing portion of this project was discontinued in 1999; therefore the GHG reduction for this project is reduced from past years.
Estimation Method	Avoided conventional auto travel was recorded for the two components of the program. In 1995, the Corporation's Ridesharing Program resulted in 205,902 avoided travel miles due to car and van pooling and 166,329 avoided miles due to the use of mass transport. The reference case assumes the mass transit services would be operational with or without this program. Consequently, if not for this program 372,231 additional vehicle miles would be traveled. This figure plus the 6800 miles of conventional auto use avoided by the Corporation's electric vehicle testing program were totaled and the Corporate Average Fuel Efficiency rate of 27.5 miles per gallon was used to calculate the motor fuel use avoided. This figure was converted to emissions data using following methodology from the Guidelines: Annual Emissions = (Annual Mileage eliminated x FMij) + (Annual Fuel saved x FFij) where FMij = emissions factor per mile eliminated for GHG i and fuel j FFij = emissions factor per unit of fuel saved for GHG i and fuel j. Fuel type was assumed to be gasoline in all cases. Emissions factors from Guidelines are as follows: FMij (grams/mile): CH_4: 0.05 CO_2: 2.0 N_2O: 0.05 FFij (grams/gallon): CH_4: 8.67 CO_2: 1.10 x 10e4 N_2O: 0.175 For all gases, FMij represents direct emissions, i.e. emission from vehicles. For N_2O and CH_4, FF_{ij} represents indirect emissions, i.e. emissions from upstream processes such as refining. For CO_2, FFij includes both tailpipe emissions (direct) and upstream emissions (indirect). A factor for direct emissions was determined using the emission factor from Appendix B in the instructions (19.641 #/gal or 0.891 x 10e4 grams/gal). A factor for indirect emissions was calculated by subtraction, 1.10 x 10e4 - 0.8909 x 10e4 = 0.209 x 10e4 grams/gal. Grams of each GHG were then converted to short tons using the following formula: (grams x 0.001) x 0.001102 (The offsetting emissions associated with the electricity used to recharge vehicle batteries were not estimated.) All employee trip reduction emissions saved are counted as indirect emissions reductions.

Los Angeles Department of Water and Power

Electric Vehicles	117 vehicles	Direct: 266.55
Project Description	DWP operates a fleet of electric vehicles (EVs): 66 for general use, 8 for carpools, 17 VIP loaners, plus 26 buses used for the San Pedro Trolley and as shuttle buses during major conventions and local events. The Electric Vehicle Program also includes vehicle service (maintenance) and deployment, infrastructure improvements (installation of charging stations for cars, buses, airport shuttles), mass transit (San Pedro Trolley, Airport Shuttle Buses, including operations, maintenance, and capital outlays), as well as public outreach & education about electric vehicles. The electric vehicle program also encourages EV use in the city and provides subsidies for infrastructure improvements related to the use of EVs.	
Estimation Method	Year 2000: Total miles saved (gasoline-powered vehicles) = 1,025,753, which includes vehicle miles eliminated due to carpooling and mass transit ridership of the electric San Pedro Trolley. (1025753 miles)(1 gal gas/20 miles)=51287.65 gals Formula No. 1: Emissions reduced from gasoline-powered vehicles = (51287.65 gallons of gasoline saved) x (19.564 lb CO_2 /gal) / 2000 lbs per ton = 501.696 tons CO_2 saved. Formula No. 2: Offsetting emissions from power plants = (253200 kWh of electricity used to charge vehicles) / 1,000,000] x (821 short tons CO_2 /GWH system emissions factor) = 207.877 tons CO_2 emitted. Formula No. 3: Net CO_2 reductions = Emissions reduced from vehicles - offsetting emissions from power plants = 501.696 tons - 207.877 tons = 293.82 tons CO_2 reduced. NOTE: System emission factors (st of CO_2 /GWH) for previous years: 1996 = 968; 1997 = 972; 1998 = 965; 1999 = 856	

National Grid USA

Electric Vehicles	2 vehicles, 8151 VMT	Direct: 2.89 Indirect: -1.80
Project Description	Massachusetts Electric Company has made available for employee use, two Toyota RAV4-EV's to demonstrate that these zero-emission vehicles are practical, and to encourage their use instead of conventional gasoline-powered cars or pick-up trucks.	
Estimation Method	The two RAV4-EVs together logged 8,151 miles from 2/16 2000 - 2/26/2001. Assumed that each vehicle would have consumed about 1 gallon of gasoline for every 25 miles driven. Utilized the Emission Coefficient from Appendix B of the Instructions for Form EIA-1605 of 19.564 pounds of CO_2 per gallon of motor gasoline. Assumed that the RAV4-EVs get 3 miles per ac kWh of charging energy. Therefore: 8,151 miles X 1 gallon MV/25 mpg = 326 gallons. For the electricity associated indirect emission: 1999: (8000 mi) / (3 mi/kWh) / (1000 kWh/MWh) * (0.729 st/MWh) * (2000 lbs/st) = 3888 lbs CO2 2000: (8151 mi) / (3 mi/kWh) / (1000 kWh/MWh) * (0.729 st/MWh) * (2000 lbs/st) = 3961 lbs CO2	

Niagara Mohawk Power Corporation		
Alternative Fuel Vehicles	Number of vehicles varied between 9 and 52;	Direct: 22.04
	30 vehicles in 2000	
Project Description	NMPC has been involved in operating and testing alternative fuel vehicles (AFVs) for almost 30 years. The Company also currently has a number of "Clean Air" natural gas-fueled buses in operation as part of a cooperative program with the Syracuse, New York Centro transit system.	
Estimation Method	CO_2 emission reductions are based on the difference in CO_2 emissions between gasoline-fueled vehicles and CNG or electric vehicles. Only direct emission reductions are reported. Emissions estimates are based on a CO_2 emission factor for each fuel. For motor gasoline, an emission factor of 19.641 lbs/gallon was used. For diesel fuel, an emission factor of 22.384 lbs/gallon was used. For CNG vehicles, a factor of 120.593 lbs/Mcf was used. These factors are based upon Form EIA-1605, Voluntary Reporting of Greenhouse Gases, Instructions, Appendix B. Fuel and Energy Source Codes and Emission Coefficients: EIA, 1996. For electric vehicles, NYPPs marginal emissions rate of 1.44 lbs/kWh for the years 1991-1995, rate of 1.48 lbs/kWh for 1996, and 1.46 lbs/kWh for 1997 and 1998 were used. These marginal rates were determined based on production simulation modeling (PROMOD IV).	

NiSource/NIPSCO	
Electric Vehicles	1 vehicle in 1994-1997	N/A
Project Description	NIPSCO expects to be a force in our region for educating the public on the environmental and efficiency benefits of electric vehicles. We lead by example with one electric truck in our fleet.
	Ozone is formed through the reaction of two precursor chemicals: Volatile Organic Compounds (VOCs) and Nitrogen Oxides (NO_x). The electric vehicle will reduce emissions of VOCs and NOX and therefore decrease the amount of ozone formed. Additionally, most of the electricity needed to serve this market will be generated during off-peak hours, thereby allowing NIPSCO to benefit from a more efficient use of generating capacity.
	NIPSCO is currently participating in or will participate on the EPRI Transportation Business Council, the Mid-America Electric Vehicle Consortium, and the Electric Transportation Consortium. In addition, we support the Electric Racing Series and the University of Notre Dame electric car and we are trial marketing "The Clean Switch" catalog.
Estimation Method	Electric Truck Purchased in 1994
	Assumptions:
	Driven 0 miles in 1998
	MGP for equivalent gas powered S-10 pickup truck = 22 mpg
	Energy used for electric power = 0.2 KWhr/mile

Calculations:

Calc. 1   (miles) x (0.2 KWhr/mile) = Energy in KWhr
Calc. 2   (KWhr)(1990 Heat rate net period)(1990 HHV Coal) = Equivalent Coal (1990 Heat rate = 10, 656 BTU/KVhr; 1990 HHV = 10,812 BTU/lb)
(KVhr) x (10,656 BTU/KVhr) / (10,812 BTU/lb) = Fuel burned in lbs.
Calc. 3   Mile x mpg = gas consumed
Calc. 4   (Equivalent Coal) x (0.6 lbs carbon/lb coal) x (3.67 lbs CO2/lb carbon) = Equivalent lbs CO2 from electric
Calc. 5   (Miles/mpg) x (19.641 lbs CO2/gallon) = Equivalent lbs CO2 from Gasoline
Calc. 6   Difference between CO2 from gasoline and electric (Coal)

There were significant repairs needed to the vehicle in 1998. NIPSCO elected not to repair the vehicle.

PG&E Corporation

Electric Vehicles	32 vehicles in 1999; 26 vehicles in 2000	Direct: 2661.68

Project Description	Pacific Gas and Electric Company Clean Air Vehicle Program:

In 1990 Pacific Gas and Electric Company received California Public Utility Commission approval to spend up to $50 million by December 31, 1994 to support the development and introduction of electric and natural gas vehicles. By the end of 1993, Pacific Gas and Electric Company was operating 698 natural gas vehicles and 30 natural gas refueling stations. Encouragement took many forms: demonstrating vehicle and station performance, providing natural gas refueling station designs, providing partial funding for vehicle purchases, opening Company stations for public use, etc. After 1994, there was a decreased emphasis on customer financial support. But the Company has continued to promote, facilitate and encourage electric and natural gas vehicle use by its customers. Pacific Gas and Electric Company continues to claim credit for not only its own fuel displacement, but also for displacements that it has encouraged its customers to undertake.

Estimation Method	Electric fuel use reflects an estimate which takes into consideration Pacific Gas and Electric Company records of its own electric vehicle fleet use, and our records of electrical energy demand within electric vehicle tariffs, manufacturer reports of electric vehicles leased or sold to our customers, and our expectations of total energy use by such vehicles.

Using the following factors, the Company calculates the CO_2 emissions avoided through displaced gasoline. Pacific Gas and Electric Company's average fossil fuel emission rate in 1998 was 0.545 tons CO_2 per MWh. The Company believes this to be the appropriate metric for generation to supply electric vehicles because fossil generation is typically at the margin in its energy mix.

0.545 tons CO_2 per MWh from Company generation facilities
19.564. lbs CO_2 per equivalent gallon of gasoline
7 KWh per gallon of gasoline in PG&E fleet

In 1999 a total of 1.05 GWh of electricity was used to displace gasoline.

1.465 tons CO_2 gasoline - 572 tons CO_2 electricity = 893 tons CO_2 avoided

An identical methodology was applied to year 2000 data.

Portland General Electric Co.

| Electric Fleet Vehicles | 76,000 VMT over 5 years; 16,000 VMT in 2000 | Direct: 4.42 |

Project Description
PGE purchased two electric vehicles in April 1996 for general fleet use.

Estimation Method
We know that 2 vehicles were converted in April of 1996. We assume the fleet vehicles travel 8000 mi/year each, that the gasoline mileage is 20 mi/gal, and that each gasoline vehicle emits 7838 pounds of CO_2 per year and each Electric vehicle emits 3895 pounds per year.

PPL Corporation

| Electric Vehicles | 8 vehicles / 10,000 VMT in 1998; 13 vehicles / 12024 VMT in 1999; 5 vehicles / 4625 VMT in 2000 | Direct: 0.85 |

Project Description
In order to foster interest in electric vehicles (EVs) PPL Corporation has established initiatives in the areas of legislation, use, and demonstration. PPL Corporation supports state and national legislation pertaining to electric vehicles. PPL Corporation owns and operates a small fleet of EVs and had 5 electric vehicles in operation in 2000. Under provisions of the Energy Policy Act of 1992, PPL Corporation began using 8 EVs in fleet operations in 1998, and added 5 more in 1999. That EV data is reported here for the 2000 reporting year.

Estimation Method
CO_2 Emissions

Conventional Gasoline Fueled Vehicle vs. Electric Vehicle (EV)

Vehicle	Number of Vehicles	Average Miles per Year	Total Miles	Grams CO_2 per mile	Tons CO_2 per year
EV	5	925	4,625	786	4.00 -- Note 1
Gasoline	5	925	4,625	970	4.94 -- Note 2

Annual Tons CO_2 saved   0.94   using the 5 electric vehicles rather than gasoline vehicles. (Derived by subtracting the tons/year of CO_2 from the two types of vehicles in the table above)

Note 1: CO_2 gm/mile, PPL 2000 Average Generation Mix of 1.103 lbs CO2/KWH using 1.46 kwh/mile at the meter and 7 percent transmission loss back to the power plant.

Note 2: CO_2 gm/mile based on passenger gasoline vehicle taken from EPA report, "Preliminary Electric Vehicle Emissions Assessment," November 3, 1993, based on 27.5 mpg and available energy of 114,000 Btu/gal gasoline. For the trucks, use 10.0 MPG instead of 27.5 MPG gives 970 gms CO_2/mile.

The Y2000 data completes PPL reporting on this project under its Climate Challenge Agreement with DOE. Emission reductions for this project may continue beyond this time, but reporting of future results will be determined by PPL on a case by case basis.

Public Utility District No. 1 of Snohomish County

Electric Car Race		N/A
Project Description	1 race in 1996	

Snohomish County PUD Sponsored a high school electric car race. Each car had two twelve volt batteries. The winner of the race was determined by who could go around an oval race track the most times in one hour.

Estimation Method	No emissions estimates were made for this project.

Sacramento Municipal Utility District

Ride Electric		Direct: 9.07
Project Description	445,452 VMT over 7 years (1994-2000)	

A key component of SMUDs Ride Electric Program is introduction and practical use of various electric vehicles in fleet service. Between 1990 and 2000, the District has acquired or assisted in acquisition of a large number of electric vehicles for its fleet and other Sacramento area fleets. These vehicles, which replaced gasoline and diesel powered vehicles, have accumulated 440,000 miles of service at SMUD, resulting in significant reductions in fossil fuel use, criteria pollutants and GHGs, to the benefit of the local community and its citizen-ratepayers.

Estimation Method

Actual Vehicle Miles Traveled (VMT) are tabulated for each vehicle in operation under the Ride Electric Program from District Vehicle Management System (VMS) records and other records maintained by the District.

Gasoline and diesel fuel displacement is calculated by multiplying VMT of the electric vehicles in the District by the fuel use (gal/mile) of an equivalent gasoline or diesel powered vehicle, or in the case of a conversion from internal combustion engine (ICE) power to electric, actual ICE fuel use figures are used for the comparison.

Emission rates for gasoline or diesel vehicles replaced by electric vehicles in the Ride Electric program are calculated by multiplying the vehicle fuel use rate (gal/mi) by the VMT by the EIA CO_2 rate for the displaced fuel, and converting from lbs. CO_2 to short tons of CO_2.

Emission rates for electric vehicles are calculated by multiplying the vehicle energy use (MWH/mi) by the vehicle miles by the CO_2 generation rate (1485 lbs CO_2/MWh) of the Districts marginal generating resources. Reductions are the net of these emissions.

Southern Company	
Transportation Research	Direct: 927
	484 vehicles
Project Description	Electric Transportation Technology - Southern Company continues to play a key role among the nation's electric utilities in the development and demonstration of energy efficient electric transportation technology. During 2000, Southern Company successfully worked to expand the penetration of material handling vehicles, Neighborhood Electric Vehicles (NEV's), and airport ground support equipment within the service territories. Electric and hybrid electric buses are now in service in Birmingham, and they are to be added in Mobile, Atlanta, and Gadsden. Fast charge projects for forklifts, airport ground support equipment, and buses are being conducted in Atlanta and Birmingham.

Southern Company completed its involvement with USABC in 2000 at the conclusion of Phase II of that effort. Through its individual corporate contribution and through its affiliation with EPRI, Southern Company has invested more that $2.1million over the eight year duration of Phases I and II of USABC. Southern Company has also provided management and technical manpower in support of USABC's efforts to develop advanced batteries for electric transportation and stationary applications.

EPRI Electric Transportation Business Unit - Southern Company is the largest individual contributor to the Electric Transportation Business Unit of EPRI. In 2000, Southern Company contributed approximately $590,000 to EPRI in support of a variety of programs to develop and demonstrate EV batteries, charging infrastructure, public transit technologies, and industrial vehicles. In 2001 Southern Company will contribute approximately $400,000 to continue this research.

The Commercialization of Electric Transportation - Southern Company supports the commercialization of EV's through internal purchases, and by supporting commercial and industrial customers in evaluating and purchasing EV's. Georgia Power Company continues an employee lease program for Southern Company employees in Atlanta by providing the opportunity to lease up to 100 EV's per year. Alabama and Georgia Power both have successful customer EV loaner programs that have caused customers to lease vehicles based on business case, even though vehicle availability from OEM's has decreased significantly. In 2000 Southern Company expanded its fleet to approximately 350 EV's, including cars, trucks, neighborhood electric vehicles, and buses.

Electric Vehicle Association of the Americas (EVAA): The mission of the EVAA is to provide a public policy framework that supports development of a widespread, sustainable market for electric vehicles. The EVAA serves as the official information source for electric vehicle technology and facilitates programs for market development. EVAA provides a national directory for electric vehicle recharging facilities and has developed infrastructure and technical reference manuals for EV Ready workshops sponsored by DOE. EVAA, with the support of Southern Company and other members, has been successful in lobbying Congress for significant dollars for airport funding and electric bus projects, as well as legislation that provides federal incentive and tax credit legislation promoting the purchase of electric vehicles. |
| **Estimation Method** | The reduction in CO2 due to operation of electric vehicles was calculated as follows:

{[(miles driven/vehicle / 22 miles/gal) x 19.564 lb CO2/gal x no. vehicles] - (MWh x coal-fired heat rate x 205.3 lb CO2/MBtu)} / 2204.6 lb/mt |

Tennessee Valley Authority	
Alternate Fuel Vehicles	N/A
	19,760 VMT (1994)

	Project Description	In 1994, TVA had 31 alternate fuel vehicles operating in its transportation fleet. These included 23 sedans fueled by M-85 (a blend of 85 percent methanol and 15 percent gasoline), 2 compressed natural gas vans, 5 electric pickup trucks, and one electric van.
		In question 4, the alternate fuel type listed as "ZZ" is the M-85.
		Project results for 1995, 1996, 1997, 1998, 1999 and 2000 are not reported as data were not available.
	Estimation Method	The direct emissions shown in Part 3 are the emissions used to compute the reported emissions reductions. These are the total emissions from the TVA transportation fleet. The actual CO_2 emissions were determined from the fuel consumed and the fuel emissions factor from Appendix B. See the previous project, Transportation Fleet Fuel Efficiency Improvements.
		The CO_2 reductions as a result of alternate fuel vehicle (AFV) operation is the net difference between the modified reference case CO_2 emissions and the actual emissions from the AFVs. The modified reference case emissions are the emissions that would have occurred had the miles driven by the AFVs been driven by the conventional fleet. The modified reference case emissions were determined from the actual AFV miles traveled, the average miles per gallon for the comparable conventional vehicles, the heating value of gasoline (125,100 BTU/Gal), and the gasoline emissions factor from Appendix B (157 lb $CO2/MM$ BTU). It was assumed that the electric and CNG vehicles displaced emissions from the conventional 4X2 pickup fleet and the M-85 vehicles displaced emissions from the conventional sedan fleet.
		The actual emissions for the CNG and M-85 AFVs were determined from the fuel usage, the heating value of the fuel, and the fuel emissions factor. The heating value for CNG is 1000 BTU/Ft3 and for M-85 is 73,590 BTU/Gal. The emissions factor for CNG is 120 lbs CO_2/MM BTU and 146 lbs CO_2/MM BTU for M-85.
		To determine the actual emissions for the electric vehicles it was assumed that the energy used to charge the vehicles was generated by the TVA coal fired system. The emissions associated with the charging was determined from the KWH used, the average coal fired system heat rate, and the coal emissions factor from Appendix B.
		The following table summarizes the operation of the AFVs and the resulting effect on CO_2 emissions for 1994. In this table, negative changes, i.e. reductions, are shown in parentheses.

Alt Fuel	Change in Miles Driven	Alt Fuel Used	Conv. Vehicle MPG	Change in Gasoline Gallon	Conv. Vehicle CO2 Tons	Heat Rate BTU/KWH	Fossil Fuel CO2 Tons	Change in CO₂ Emission Tons
M-85	14258	544 Gal	29.8	(478)	(4.7)	--	2.9	(1.8)
CNG	1301	25000 CF	15.5	(84)	(0.8)	--	1.5	0.7
Elec.	4201	1360 KWH	21.2	(198)	(1.9)	10047	1.4	(0.5)
TOTAL	19760			(760)	(7.5)		5.8	(1.6)

Waverly Light & Power Company		
Electric Vehicle Project	1 vehicle (1992-1999)	N/A
Project Description	WLP converted a line truck to an all electric vehicle in 1992. The unit offsets CO_2 emission from gasoline, although these offsets are not cumulative in nature. The truck was removed from service in 2000.	
Estimation Method	The CO_2 reductions are all based upon engineering estimates for both total miles driven and the approximate gas mileage for the vehicle. The analysis assumes that electricity for operating the truck was generated by Waverly hydroelectric generating capability. Estimated Mileage/Year: 3,000 miles Assumed Miles/Gallon: 22 mpg CO_2 Emission Factor from Gasoline: 19.65 lb CO_2/gal Year / Fuel Usage / Year (gal) / CO_2 Reduction (TPY) 1992 136.4 1.3 1993 136.4 1.3 1994 136.4 1.3 1995 136.4 1.3 1996 136.4 1.3 1997 136.4 1.3 1998 136.4 1.3 1999 136.4 1.3 2000 0 0 The vehicle was removed from service in 2000.	

Project Size refers to size in 2000 unless otherwise noted. VMT = vehicle miles traveled

^ Project Description and Estimation Method are quoted directly from the Reporters' 2000 EIA-1605 reports.

[this page deliberately left blank]

A5

U.S. Initiative on Joint Implementation (USIJI) Project Criteria

Criteria from the Final USIJI groundrules as published in the Federal Register on June 1, 1994:

"Section V—Criteria

A.   To be included in the USIJI, the Evaluation Panel must find that a project submission:

(1)   Is acceptable to the government of the host country;

(2)   Involves specific measures to reduce or sequester greenhouse gas emissions initiated as the result of the U.S. Initiative on Joint Implementation, or in reasonable anticipation thereof;

(3)   Provides data and methodological information sufficient to establish a baseline of current and future greenhouse gas emissions:

   (a)   In the absence of the specific measures referred to in A.(2)-- of this section; and

   (b)   As the result of the specific measures referred to in A.(2) of this section;

(4)   Will reduce or sequester GHG emissions beyond those referred to in A.(3)(a) of this section, and if federally funded, is or will be undertaken with funds in excess of those available for such activities in fiscal year 1993;

(5)   Contains adequate provisions for tracking the GHG emissions reduced or sequestered resulting from the project, and on a periodic basis, for modifying such estimates and for comparing actual results with those originally projected;

(6)   Contains adequate provisions for external verification of the greenhouse gas emissions reduced or sequestered by the project;

(7)   Identifies any associated non-greenhouse gas environmental impacts/benefits;

(8)   Provides adequate assurance that greenhouse gas emissions reduced or sequestered over time will not be lost or reversed; and

Provides for annual reports to the Evaluation Panel on the emissions reduced or sequestered, and on the share of such emissions attributed to each of the participants, domestic and foreign, pursuant to the terms of voluntary agreements among project participants.

B.　　In determining whether to include projects under the USIJI, the Evaluation Panel shall also consider:

　　　　(1)　　The potential for the project to lead to changes in greenhouse gas emissions elsewhere;

　　　　(2)　　The potential positive and negative effects of the project apart from its effect on greenhouse gas emissions reduced or sequestered;

Whether the U.S. participants are emitters of GHGs within the United States and, if so, whether they are taking measures to reduce or sequester such emissions; and

Whether efforts are underway within the host country to ratify or accede to the United Nations Framework Convention on Climate Change, to develop a national inventory and/or baseline of greenhouse gas emissions by sources and removals by sinks, and whether the host country is taking measures to reduce its emissions and enhance its sinks and reservoirs of greenhouse gases."

A6

U.S. Department of Energy State Average Electricity Emission Factors[163]

Table A6	State Average Electricity Emission Factors				
	Carbon Dioxide			Methane	Nitrous Oxide
Region/State	lbs/kWh	short tons/ MWh	metric tons/ MWh	lbs/MWh	lbs/MWh
New England	**0.98**	**0.491**	**0.446**	**0.0207**	**0.0146**
Connecticut	0.94	0.471	0.427	0.0174	0.0120
Maine	0.85	0.426	0.386	0.0565	0.0270
Massachusetts	1.28	0.639	0.579	0.0174	0.0159
New Hampshire	0.68	0.341	0.310	0.0172	0.0141
Rhode Island	1.05	0.526	0.477	0.0068	0.0047
Vermont	0.03	0.014	0.013	0.0096	0.0039
Mid Atlantic	**1.04**	**0.520**	**0.471**	**0.0093**	**0.0145**
New Jersey	0.71	0.353	0.320	0.0077	0.0079
New York	0.86	0.429	0.389	0.0081	0.0089
Pennsylvania	1.26	0.632	0.574	0.0107	0.0203
East-North Central	**1.63**	**0.815**	**0.740**	**0.0123**	**0.0257**
Illinois	1.16	0.582	0.528	0.0082	0.0180
Indiana	2.08	1.038	0.942	0.0143	0.0323
Michigan	1.58	0.790	0.717	0.0146	0.0250
Ohio	1.80	0.900	0.817	0.0130	0.0288
Wisconsin	1.64	0.821	0.745	0.0138	0.0260
West-North Central	**1.73**	**0.864**	**0.784**	**0.0127**	**0.0269**
Iowa	1.88	0.941	0.854	0.0138	0.0298
Kansas	1.68	0.842	0.764	0.0112	0.0254
Minnesota	1.52	0.762	0.691	0.0157	0.0247
Missouri	1.84	0.920	0.835	0.0126	0.0288
Nebraska	1.40	0.700	0.635	0.0095	0.0219
North Dakota	2.24	1.121	1.017	0.0147	0.0339
South Dakota	0.80	0.399	0.362	0.0053	0.0121

---

[163] Energy Information Administration, *Updated State- and Regional-level Greenhouse Gas Emission Factors for Electricity* (March 2002), http://www.eia.doe.gov/oiaf/1605/e-factor.html.

South Atlantic	**1.35**	**0.674**	**0.612**	**0.0127**	**0.0207**
Delaware	1.83	0.915	0.830	0.0123	0.0227
Florida	1.39	0.697	0.632	0.0150	0.0180
Georgia	1.37	0.683	0.619	0.0129	0.0226
Maryland*	1.37	0.683	0.620	0.0118	0.0206
North Carolina	1.24	0.621	0.563	0.0105	0.0203
South Carolina	0.83	0.417	0.378	0.0091	0.0145
Virginia	1.16	0.582	0.528	0.0137	0.0192
West Virginia	1.98	0.988	0.897	0.0137	0.0316
East-South Central	**1.49**	**0.746**	**0.677**	**0.0128**	**0.0240**
Alabama	1.31	0.656	0.595	0.0137	0.0223
Kentucky	2.01	1.004	0.911	0.0140	0.0321
Mississippi	1.29	0.647	0.587	0.0132	0.0165
Tennessee	1.30	0.648	0.588	0.0105	0.0212
West-South Central	**1.43**	**0.714**	**0.648**	**0.0087**	**0.0153**
Arkansas	1.29	0.643	0.584	0.0125	0.0203
Louisiana	1.18	0.589	0.534	0.0094	0.0112
Oklahoma	1.72	0.861	0.781	0.0110	0.0223
Texas	1.46	0.732	0.664	0.0077	0.0146
Mountain	**1.56**	**0.781**	**0.709**	**0.0108**	**0.0236**
Arizona	1.05	0.525	0.476	0.0068	0.0154
Colorado	1.93	0.963	0.873	0.0127	0.0289
Idaho	0.03	0.014	0.013	0.0080	0.0033
Montana	1.43	0.717	0.650	0.0108	0.0227
Nevada	1.52	0.759	0.688	0.0090	0.0195
New Mexico	2.02	1.009	0.915	0.0131	0.0296
Utah	1.93	0.967	0.878	0.0134	0.0308
Wyoming	2.15	1.073	0.973	0.0147	0.0338
Pacific Contiguous	**0.45**	**0.224**	**0.203**	**0.0053**	**0.0037**
California	0.61	0.303	0.275	0.0067	0.0037
Oregon	0.28	0.141	0.127	0.0033	0.0034
Washington	0.25	0.123	0.111	0.0037	0.0040
Pacific Non-contiguous	**1.56**	**0.780**	**0.707**	**0.0161**	**0.0149**
Alaska	1.38	0.690	0.626	0.0068	0.0089
Hawaii	1.66	0.831	0.754	0.0214	0.0183
United States	**1.34**	**0.668**	**0.606**	**0.0111**	**0.0192**

Note: These state- and regional-level electricity emission factors represent average emissions per kWh or MWh generated by utility and nonutility electric generators for the 1998-2000 time period. The Voluntary Reporting of Greenhouse Gases Program believes these factors provide reasonably accurate default values for power generated in a given state or region (U.S. Census Division). However, reporters should use these state- and regional-level factors only if utility-specific or power pool-specific emission factors are not available.

*Includes the District of Columbia

A6 State Average Electricity Emission Factors

A7 Fuel and Energy Source Emission Coefficients[164]

Table A7	Fuel and Energy Source Emission Coefficients		
	Emission Coefficients		
Fuel	**Pounds CO$_2$ per unit volume or mass**		**Pounds CO$_2$ per million Btu**
Petroleum Products			
Aviation Gasoline	18.355 770.916	per gallon per barrel	152.717
Distillate Fuel (No. 1, No. 2, No. 4 Fuel Oil and Diesel)	22.384 940.109	per gallon per barrel	161.386
Jet Fuel	21.095 885.98	per gallon per barrel	156.258
Kerosene	21.537 904.565	per gallon per barrel	159.535
Liquified Petroleum Gases (LPG)	12.805 537.804	per gallon per barrel	139.039
Motor Gasoline	19.564 822.944	per gallon per barrel	156.425
Petroleum Coke	32.397 1356.461 6768.667	per gallon per barrel per short ton	225.130
Residual Fuel (No. 5 and No. 6 Fuel Oil)	26.033 1,093.384	per gallon per barrel	173.906
Methane	116.376	per 1000 ft^3	115.258
Landfill Gas	a	per 1000 ft^3	115.258
Flare Gas	133.759	per 1000 ft^3	120.721
Natural Gas (Pipeline)	120.593	per 1000 ft^3	117.080
Propane	12.669 532.085	per gallon per barrel	139.178

---

[164] Instructions for Form EIA-1505: Voluntary Reporting of Greenhouse Gases (for data through 2001). EIA Energy Information Administration, US Department of Energy. February 2002.

Electricity	Varies depending on fuel used to generate electricity[b]		
Electricity Generated from Landfill Gas	Varies depending on heat rate of the power generating facility		
Coal			
Anthracite	3,852.16	per short ton	227.400
Bituminous	4,931.30	per short ton	205.300
Subbituminous	3,715.90	per short ton	212.700
Lignite	2,791.60	per short ton	215.400
Renewable Sources			
Biomass	Varies depending on the composition of the biomass		
Geothermal Energy	0		0
Wind	0		0
Photovoltaic and Solar Thermal	0		0
Hydropower	0		0
Tires/Tire-Derived Fuel	6160	per short ton	189.538
Wood and Wood Waste [c,d]	3120	per short ton	195.000
Municipal Solid Waste [e]	1999	per short ton	199.854
Nuclear	0		0
Other	-		-

[a] For a landfill gas coefficient per thousand standard cubic foot, multiply the methane factor by the share of the landfill gas that is methane.

[b] For average electric power emission coefficients by state, see Appendix V (Previous Page).

[c] For as-fired dry wood

[d] Wood and wood waste contain "biogenic" carbon. Under international GHG accounting methods developed by the Intergovernmental Panel on Climate Change, biogenic carbon is considered to be part of the natural carbon balance and does not add to atmospheric concentrations of carbon dioxide.[165] Reporters may wish to use an emission factor of zero for wood, wood waste, and other biomass fuels in which the carbon is entirely biogenic.

---

[165] Intergovernmental Panel on Climate Change. *Greenhouse Gas Inventory Reference Manual: Revised 1996 IPCC Guidelines for National Greenhouse Gas Inventories*, Vol. 3, Pg. 6.28, (Paris France 1997).

A7 Fuel and Energy Source Emission Coefficients

References

AIJ Uniform Reporting Format: Activities Implemented Jointly under the Pilot Phase. *The RABA/IKARUS Compressed Natural Gas Engine Project.* http://www.unfccc.int/program/aij/aijact/hunnld01.html.

American Automobile Manufacturers Association (AAMA). "World Motor Vehicle Data 1993." AAMA, 1993.

American Automobile Manufacturers Association (AAMA). "Motor Vehicle Facts and Figures 1996." AAMA, 1996.

Austin, T., Dulla, R. and Carlson, T. *Alternative and Future Fuels and Energy Sources For Road Vehicles.* Sierra Research Inc, 8 July 1999. http://www.tc.gc.ca/envaffairs/subgroups/vehicle_technology/study2/Final_r port/Final_Report.htm.

Bunch, D.S., et al., *Demand for Clean-Fuel Personal Vehicles in California: A Discrete-*

University of California at Irvine, Institute of Transportation Studies, *Choice Stated Preference Survey.* Report UCT-ITS-WP-91-8, 1991.

California Air Resources Board, "California Exhaust Emissions Standards and Test Procedures for 2001 and Subsequent Model Passenger Cars, Light-Duty Trucks, and Medium-Duty Vehicles." Proposed Amendments, 28 September 2001.

California Air Resource Board. *Low-Emission Vehicle Program.* 28 September 2001. http://www.arb.ca.gov/msprog/levprog/levprog.htm.

California Air Resources Board. *California's Zero Emission Vehicle Program Fact Sheet.* 6 December 2001. http://www.arb.ca.gov/msprog/zevprog/factsheet/evfacts.pdf.

California Air Resources Board. "Notice of Public Hearing to Consider the Adoption of Amendments to the Low-Emission Vehicle Regulations." The California Low-Emission Vehicle Regulations, 15 November 2001.

California Energy Commission, *Global Climate Change and California.* http://www.energy.ca.gov/global_climate_change/index.html.

Center for Biological Diversity v. Abraham, N.D. Cal., No. CV-00027. 2 January 2002.

Cuenca, R.M., Gaines, L.L., and Vyas, A.D. *Evaluation of Electric Vehicle Production and Operating Costs.* Argonne National Laboratory, November 1999.

EarthVision Environmental News. "New York Adopts New California Emission Standards." November 2002. http://www.climateark.org/articles/2000/4th/nyadnewc.htm.

Electric Vehicle Association of the Americas (EVAA), *State Laws and Regulations Impacting Electric Vehicles.* January 2002. http://www.evaa.org.

Energy Information Administration, *Instructions for Form EIA-1505: Voluntary Reporting of Greenhouse Gases (for Data through 2001)*. February 2002.

Energy Information Administration, *Updated State- and Regional-level Greenhouse Gas Emission Factors for Electricity*. March 2002. http://www.eia.doe.gov/oiaf/1605/e-factor.html.

Federal Register, "Electric Vehicles." Volume 61, Number 51, pages 10627-10628.

Fulton, L,, Liliu, C., Landwehr, M., and Schipper, L. *Saving Oil and Reducing CO_2 Emissions in Transport: Options and Strategies*. International Energy Agency (IEA), 2001.

Intergovernmental Panel on Climate Change. *Greenhouse Gas Inventory Reference Manual: Revised 1996 IPCC Guidelines for National Greenhouse Gas Inventories Vol. 3*. 1997.

International Energy Agency (IEA) and Organisation for Economic Cooperation and Development (OECD). *Good Practice Greenhouse Abatement Policies: Transport Sector*. OECD and IEA Information Papers prepared for the Annex I Expert Group on the UNFCCC. International Energy Agency (IEA) and Organisation for Economic Cooperation and Development (OECD), November 2000.

Maryland Green Buildings Council, "2001 Green Buildings Council Report." November 2001. http://www.dgs.state.md.us/GreenBuildings/Documents/FullReport.pdf.

Massachusetts Low Emission Vehicle Program. Public Hearings on the Amendments to the State Implementation Plan for Ozone and Hearing and Findings under the Massachusetts Low Emission Vehicle Statute - 310 CMR 7.40. February 2002. http://www.state.ma.us/dep/bwp/daqc/daqcpubs.htm.

National Oceanic and Atmospheric Administration. *Clear Skies & Global Climate Change Initiatives*. 14 February 2002. http://www.whitehouse.gov/news/releases/2002/02/20020214-5.html.

Office of the Governor of New York, *Regulation to Reduce Harmful Vehicle Emissions, Alternative to Promote Clean Vehicle Technology, Improve Air Quality*. 4 January 2002. http://www.state.ny.us/governor/press/year02/jan4_02.htm.

Rosenzweig, R., Varilek, M., Feldman, B., Kuppalli, R. and Jansen, J. *The Emerging International Greenhouse Gas Market*. Pew Center on Global Climate Change. March 2002.

State of Maine Department of Environmental Protection. *Rule Chapter 127, New Motor Vehicle Emission Standard*. Basis Statement for Amendments of 21 December 2000.

Tompkins, M., et al. *Determinants of Alternative Fuel Vehicle Choice in the Continental United States*. Transportation Research Record No. 1641. Transportation Research Board, 1998.

University of California at Irvine, Institute of Transportation Studies, *Choice Stated Preference Survey*. Report UCT-ITS-WP-91-8, 1991.

U.S. Department of Energy, *Alternative Fuel Vehicle Fleet Buyer's Guide*. http://www.fleets.doe.gov/cgi-in/fleet/main.cgi?17357,state_ins_rep,5,468050.

U.S. Department of Energy, *Federal Fleet AFV Program Status*. 2 June 1998. http://www.ccities.doe.gov/pdfs/slezak.pdf.

References

U.S. Department of Energy, Office of Energy Efficiency and Renewable Energy. *Just the Basics: Electric Vehicles, Transportation for the 21st Century.* January 2002.

U.S. Environmental Protection Agency, *Legislative Initiatives.* http://yosemite.epa.gov/globalwarming/ghg.nsf/actions/Legislative Initiatives.

Wang, M. *Regulated Emissions and Energy Use in Transportation (GREET).* Argonne National Laboratory. www.transportation.anl.gov/ttrdc/greet.

White House Office of the Press Secretary. "President Announces Clear Skies & Global Climate Change Initiatives." 14 February 2002. http://www.whitehouse.gov/news/releases/2002/02/20020214-5.html.

White House. Global Climate Change Policy Book. February 2002. http://www.whitehouse.gov/news/releases/2002/02/climatechange.html.

World Resources Institute (WRI). *Proceed With Caution: Growth in the Global Motor Vehicle Fleet.* http://www.wri.org/trends/autos.html.

World Resources Institute (WRI) and World Business Council for Sustainable Development (WBCSD). *The Greenhouse Gas Protocol: A Corporate Accounting and Reporting Standard.* www.GHGprotocol.org.

World Resources Institute (WRI) and World Business Council on Sustainable Development (WBCSD) Greenhouse Gas (GHG) Protocol Initiative. *Calculating CO_2 Emissions from Mobile Sources.* www.GHGprotocol.org.